ENVIRONMENTAL RESEARCH ADVANCES

WETLANDS

OVERVIEW AND TARGETED INVESTMENT FOR RESTORATION AND PROTECTION

ENVIRONMENTAL RESEARCH ADVANCES

Additional books in this series can be found on Nova's website under the Series tab.

Additional e-books in this series can be found on Nova's website under the e-book tab.

ENVIRONMENTAL RESEARCH ADVANCES

WETLANDS

OVERVIEW AND TARGETED INVESTMENT FOR RESTORATION AND PROTECTION

NELSON E. SANTIAGO
EDITOR

nova publishers
New York

NOTICE TO THE READER

Library of Congress Cataloging-in-Publication Data

ISBN: 978-1-63482-728-7

Published by Nova Science Publishers, Inc. † New York

CONTENTS

PREFACE

Wetlands, with a variety of physical characteristics, are found throughout the country. They are known in different regions as swamps, marshes, fens, potholes, playa lakes, or bogs. Although these places can differ greatly, they all have distinctive plant and animal assemblages because of the wetness of the soil. Some wetland areas may be continuously inundated by water, while other areas may not be flooded at all. In coastal areas, flooding may occur daily as tides rise and fall. Recent Congresses have considered numerous policy topics that involve wetlands. Many reflect issues of long-standing interest, such as applying federal regulations on private lands, wetland loss rates, and restoration and creation accomplishments. This book provides an overview of issues with the wetlands; and provides some economic insights on targeting investments to cost effectively restore and protect wetland ecosystems.

Chapter 1 - Recent Congresses have considered numerous policy topics that involve wetlands. Many reflect issues of long-standing interest, such as applying federal regulations on private lands, wetland loss rates, and restoration and creation accomplishments.

The issue receiving the greatest attention recently has been determining which wetlands should be included and excluded from requirements of the Clean Water Act (CWA), especially the Section 404 permit program that regulates waste discharges affecting wetlands, which is administered by the Army Corps of Engineers and the Environmental Protection Agency (EPA). As a result of Supreme Court rulings in 2001and 2006 that narrowed federal regulatory jurisdiction over certain isolated wetlands, the jurisdictional reach of the permit program has also been narrowed. In response, legislation intended to reverse the Court's rulings in these cases has been introduced in Congress but not enacted. In March 2014, EPA and the Army Corps proposed a rule to define the scope of waters protected by the CWA. According to the agencies, the proposed rule would revise the existing administrative definition of "waters of the United States" consistent with the Supreme Court's rulings and consistent with science concerning the interconnectedness of tributaries, wetlands, and other waters and the effects of these connections on the chemical, physical, and biological integrity of downstream waters. The proposal, which could be finalized in 2015, is controversial with groups who contend that it would vastly increase federal assertion of jurisdiction that triggers CWA regulatory requirements.

Wetland protection efforts continue to engender controversy over issues of science and policy. Topics include the rate and pattern of loss, whether all wetlands should be protected in a single fashion, the effectiveness of the current suite of laws in protecting them, and the fact that 75% of remaining U.S. wetlands are located on private lands.

Many public and private efforts have sought to mitigate damage to wetlands and to protect them through acquisition, restoration, enhancement, and creation, particularly coastal wetlands. While recent data indicate success in some restoration efforts, leading to increases in some types of wetlands in some locations, many scientists question if restored or created wetlands provide equivalent replacement for natural wetlands that contribute multiple environmental services and values.

One reason for controversies about wetlands is that they occur in a wide variety of physical forms, and the numerous values they provide, such as wildlife habitat, also vary widely. In addition, the total wetland acreage in the lower 48 states is estimated to have declined from more than 220 million acres three centuries ago to 110.1 million acres in 2009. The national policy goal of no net loss, endorsed by administrations for the past two decades, had been reached by 2004, according to the Fish and Wildlife Service, as the rate of loss had been more than offset by net gains through expanded restoration efforts authorized in multiple laws. However, more recent data show wetlands losses of nearly 14,000 acres per year. Many protection advocates say that gains do not necessarily account for the changes in quality of the remaining wetlands, and many also view federal protection efforts as inadequate or uncoordinated. Others, who advocate the rights of property owners and development interests, characterize these efforts as too intrusive. Numerous state and local wetland programs add to the complexity of the protection effort.

Chapter 2 - USDA has spent more than $4.2 billion on wetland restoration and protection over the last two decades. One challenge in allocating these funds is the lack of information on variations in wetland benefits and costs across the Nation. This report discusses the biophysical impacts of new wetlands for eight benefit categories: duck hunting, carbon sequestration, flood protection, nitrogen removal, species protection, open space, sediment removal, and groundwater recharge, as well as the value of these impacts for some categories. In addition, it presents county-level estimates of the costs of restoring and preserving wetlands for some parts of the United States. Although the estimates range in precision and are not comprehensive, they call attention to some areas where the benefits of new wetlands are likely to exceed costs or perhaps may be insignificant. For example, the benefits of restoring and preserving wetlands near the Missouri River in central North and South Dakota are likely to exceed costs. Findings underscore the need for additional information that may increase the number, accuracy, and spatial resolution of wetland benefit estimates.

In: Wetlands
Editor: Nelson E. Santiago

ISBN: 978-1-63482-728-7

Chapter 1

WETLANDS: AN OVERVIEW OF ISSUES*

Claudia Copeland

SUMMARY

Recent Congresses have considered numerous policy topics that involve wetlands. Many reflect issues of long-standing interest, such as applying federal regulations on private lands, wetland loss rates, and restoration and creation accomplishments.

The issue receiving the greatest attention recently has been determining which wetlands should be included and excluded from requirements of the Clean Water Act (CWA), especially the Section 404 permit program that regulates waste discharges affecting wetlands, which is administered by the Army Corps of Engineers and the Environmental Protection Agency (EPA). As a result of Supreme Court rulings in 2001and 2006 that narrowed federal regulatory jurisdiction over certain isolated wetlands, the jurisdictional reach of the permit program has also been narrowed. In response, legislation intended to reverse the Court's rulings in these cases has been introduced in Congress but not enacted. In March 2014, EPA and the Army Corps proposed a rule to define the scope of waters protected by the CWA. According to the agencies, the proposed rule would revise the existing administrative definition of "waters of the United States" consistent with the Supreme Court's rulings and consistent with science concerning the interconnectedness of

* This is an edited, reformatted and augmented version of a Congressional Research Service publication RL33483, prepared for Members and Committees of Congress dated November 3, 2014.

tributaries, wetlands, and other waters and the effects of these connections on the chemical, physical, and biological integrity of downstream waters. The proposal, which could be finalized in 2015, is controversial with groups who contend that it would vastly increase federal assertion of jurisdiction that triggers CWA regulatory requirements.

Wetland protection efforts continue to engender controversy over issues of science and policy. Topics include the rate and pattern of loss, whether all wetlands should be protected in a single fashion, the effectiveness of the current suite of laws in protecting them, and the fact that 75% of remaining U.S. wetlands are located on private lands.

Many public and private efforts have sought to mitigate damage to wetlands and to protect them through acquisition, restoration, enhancement, and creation, particularly coastal wetlands. While recent data indicate success in some restoration efforts, leading to increases in some types of wetlands in some locations, many scientists question if restored or created wetlands provide equivalent replacement for natural wetlands that contribute multiple environmental services and values.

One reason for controversies about wetlands is that they occur in a wide variety of physical forms, and the numerous values they provide, such as wildlife habitat, also vary widely. In addition, the total wetland acreage in the lower 48 states is estimated to have declined from more than 220 million acres three centuries ago to 110.1 million acres in 2009. The national policy goal of no net loss, endorsed by administrations for the past two decades, had been reached by 2004, according to the Fish and Wildlife Service, as the rate of loss had been more than offset by net gains through expanded restoration efforts authorized in multiple laws. However, more recent data show wetlands losses of nearly 14,000 acres per year. Many protection advocates say that gains do not necessarily account for the changes in quality of the remaining wetlands, and many also view federal protection efforts as inadequate or uncoordinated. Others, who advocate the rights of property owners and development interests, characterize these efforts as too intrusive. Numerous state and local wetland programs add to the complexity of the protection effort.

INTRODUCTION

Wetlands, with a variety of physical characteristics, are found throughout the country. They are known in different regions as swamps, marshes, fens, potholes, playa lakes, or bogs. Although these places can differ greatly, they all have distinctive plant and animal assemblages because of the wetness of the soil. Some wetland areas may be continuously inundated by water, while other areas may not be flooded at all. In coastal areas, flooding may occur daily as tides rise and fall.

Prior to the mid-1980s, federal laws and policies to protect wetlands were generally limited to providing habitat for migratory waterfowl, especially ducks and geese. Some laws encouraged destruction of wetland areas, including selected provisions in the federal tax code, public works legislation, and farm programs.

Since the mid-1980s, the values of wetlands have been recognized in different ways in numerous national policies, and federal laws either encourage wetland protection, or prohibit or do not support their destruction. These laws, however, do not add up to a fully consistent or comprehensive national approach. The central federal regulatory program, found in Section 404 of the Clean Water Act, requires permits for the discharge of dredged or fill materials into many but not all wetland areas. However, other activities that may adversely affect wetlands do not require permits, and some places that scientists define as wetlands are exempt from this permit program because of physical characteristics or the type of activity that takes place. One agricultural program, swampbuster, is a disincentive program that indirectly protects wetlands by making farmers who drain wetlands ineligible for federal farm program benefits; those who do not receive these benefits (62% of all farmers received no direct payments from the farm subsidy program in 2007) have no reason to observe the requirements of this program. Numerous other acquisition, protection, and restoration programs complete the current federal effort.

Although numerous wetland protection bills have been introduced in recent Congresses, the most significant new wetlands legislation to be enacted has been in farm bills. Congress also has reauthorized several wetlands programs, mostly setting higher appropriations ceilings, without making significant shifts in policy. The George W. Bush Administration endorsed wetland protection in legislation, such as the farm bill and the North American Wetlands Conservation Act reauthorization, and at events, such as Earth Day presentations. The Bush Administration also issued rules on mitigation policies. The Obama Administration has proposed controversial changes to regulatory program jurisdiction (see discussion below).

Congress has provided a forum in numerous hearings where conflicting interests in wetland issues have been debated. These debates encompass disparate scientific and programmatic questions and conflicting views of the role of government where private property is involved. Broadly speaking, the conflicts are between:

- Environmental interests and wetland protection advocates who have been pressing for greater wetlands protection as multiple values have been more widely recognized, by improving coordination and consistency among agencies and levels of governments, and strengthened programs; and
- Others, including landowners, farmers, and small businessmen, who counter that protection efforts have gone too far, by aggressively regulating privately owned wet areas that provide few wetland values. They have been especially critical of the U.S. Army Corps of Engineers (Corps) and the U.S. Environmental Protection Agency (EPA), asserting that they administer the Section 404 program in an overzealous and inflexible manner.

Recent wetland legislative activity in Congress centered broadly on two issues. One was on wetlands conservation provisions in the farm bill. Provisions enacted in the 2014 farm bill (P.L. 113-79) repealed the wetlands reserve program, but it created a new conservation easement program, which continues voluntary efforts to protect and restore wetlands. The law also reauthorized the Conservation Reserve Program (CRP), the largest federal, private-land retirement program, but reduced the acreage enrollment cap by 25%. CRP compensates landowners who voluntarily remove land from agricultural production for the benefit of soil and water quality and wildlife habitat.

The second major area of recent legislative interest has been proposals addressing the scope of geographic jurisdiction of wetlands regulations under the Clean Water Act. This interest arises because federal courts have played a key role in interpreting and clarifying the limits of federal jurisdiction to regulate activities that affect "waters of the United States," including wetlands, especially since a 2001 Supreme Court ruling in the so-called *SWANCC* decision and another in 2006 in *Rapanos vs. United States.* For several Congresses, legislation intended to reverse the *SWANCC* and *Rapanos* rulings was introduced but not enacted. Legislation that instead would narrow the definition of "waters of the United States" also has been introduced, such as S. 890/H.R. 3377 in the 113th Congress.

WETLANDS: SCIENCE AND INFORMATION

Scientific questions about wetlands, with answers that can be important to policy makers, include how to define wetlands; how to catalogue the rate and pattern of wetland declines and losses as well as restorations and increases; and how to assess the importance of wetland changes to broader ecosystems. Wetlands science has made considerable strides in developing a fuller and more sophisticated knowledge about many aspects of wetlands in the more than two decades since protecting wetlands became a general policy goal in federal law and program administration.[1]

There are two topics where scientific information and wetland protection policies remain inconsistent: should all regulated wetlands be treated equally; and if all scientifically defined wetlands are not covered by the federal regulatory program, what subset should be covered, and how should such decisions be made? While discussion of either question has major science elements, both are primarily addressed in the section below about the Clean Water Act Section 404 program.

What Is a Wetland?

Scientists generally agree that the presence of a wetland can be determined by a combination of soils, plants, and hydrology. The only definition of wetlands in law, in the swampbuster provisions of farm legislation (P.L. 99-198) and in the Emergency Wetlands Resources Act of 1986 (P.L. 99-645), lists those three components. This definition does not include more specific criteria, such as exactly what conditions must be present and for how long, thus leaving interpretation to scientists and regulators on a case-by-case basis. Controversies are exacerbated when many sites that have those three components and are identified as wetlands by experts, either may have wetland characteristics only some portion of the time, or may not look like what many people visualize as wetlands. Also, many of these sites have been directly or indirectly modified by human activities that diminish their appearance (and their ability to perform wetland functions).

Wetlands currently subject to federal regulation are a large subset of all places that members of the scientific community would call a wetland. These regulated wetlands, under the Section 404 program discussed below, are currently identified using technical criteria in a wetland delineation manual issued by the Corps in 1987. This manual was prepared jointly and is used by all federal agencies to carry out their responsibilities under this program (the Corps, EPA, Fish and Wildlife Service [FWS], and the National Marine Fisheries Service [NMFS]). It provides guidance and field-level consistency for the agencies that have roles in wetland regulatory protection. (A second and slightly different manual, agreed to by the Corps and the Natural Resources Conservation Service [NRCS], is used for delineating wetlands on agricultural lands.) While the agencies try to improve the objectivity and consistency of wetland identification and delineation, judgment continues to play a role and can lead to site-specific controversies. Cases discussed below (see "Section 404 Judicial Proceedings: *SWANCC* and *Rapanos*") center on whether wetlands should be included or exempted from the regulatory program in certain circumstances, such as the physical setting.

What Functional Values Are Provided by Wetlands?

Functional values, both ecological and economic, at each wetland depend on its location, size, and relationship to adjacent land and water areas. Many of these values have been recognized only recently. Historically, many federal programs encouraged wetlands to be drained or altered because they were seen as having little value as wetlands (for example, flood protection programs of the Corps and Department of Agriculture have modified or eliminated many floodplain wetlands through alterations of the hydraulic/hydrologic regime). Wetland values can include

- habitat for aquatic birds and other animals and plants, including numerous threatened and endangered species; production of fish and shellfish;
- water storage, including mitigating the effects of floods and droughts;
- water purification;
- recreation;
- timber production;
- food production;
- education and research; and
- open space and aesthetic values.

Usually wetlands provide some combination of these values; single wetlands rarely provide all of these values. The composite value typically declines when wetlands are altered. In addition, the effects of alteration often extend well beyond the immediate area, because wetlands are usually part of a larger water system. For example, conversion of wetlands to urban uses has increased flood damages; this value has received considerable attention as the costs of natural disaster costs mounted since the 1990s.

How Fast Are Wetlands Disappearing, and How Many Acres Are Left?

A number of reports document changes in wetland acres. The U.S. Fish and Wildlife Service (FWS) periodically surveys national net trends in wetland acreage using the National Wetlands Inventory (NWI). It has estimated that when European settlers first arrived, wetland acreage in the area that would become the 48 states was more than 220 million acres, or about 5% of the total land area. According to its most recent report of national trends, issued in 2011, total wetland acreage in 2009 was estimated to be 110.1 million acres.[2] Until recently, NWI data had shown small annual gains overall in wetland acreage. However, the 2009 total was a slight decline in acreage over the previous five years (62,300 acres), or about 13,800 acres lost per year, reflecting a combination of some losses and some gains in acres and types of wetlands across the country. FWS also has published reports on wetland status and trends in several individual regions and states, such as Florida, Texas, Delaware, South Carolina, and Alaska.[3]

Of particular interest to scientists and natural resource managers are coastal wetlands, which provide important ecosystems services, because they serve as buffers to protect coastal areas from storm damage and sea level rise, while providing habitat for fish, shellfish, and wildlife that are commercially and recreationally important. Coastal watersheds, where these wetlands are located, are affected by population growth more than non-coastal areas, since 52% of the total U.S. population lives in counties that drain to coastal watersheds, although these counties are less than 20% of U.S. land area, excluding Alaska. Coastal wetlands are vulnerable to direct and indirect effects of residential and commercial development, pollutant discharges, and other human activities. A 2013 report by the FWS and National Oceanic and Atmospheric Administration (NOAA) found that in 2009 there were an estimated 41.1 million acres of wetlands in the coastal watersheds of the United States, representing 37.3% of total wetland area in the lower 48 states. The report also found that U.S. coastal wetlands are vanishing at a rate of more than 80,000 acres per year, about six times greater than the estimated rate of wetland loss for the entire United States. The increased loss, measured between 2004 and 2009, was attributed to severe weather in the Gulf of Mexico and urban and rural development in other areas, and the reported loss was 25% greater than the annual loss rate found in a previous report covering the years 1998 to 2004. The largest loss, according to the report, was in the Gulf of Mexico region, where 257,150 acres of coastal wetlands disappeared due to erosion and/or inundation. Throughout the Gulf region, saltwater wetlands have been adversely affected by the cumulative effect of oil and gas development that increased their vulnerability to intense storms.[4]

A study of national wetland condition by the EPA together with states, tribes, and other federal partners is underway and is expected to provide information on both quality and quantity of wetlands in the United States.[5]

In 2002, the George W. Bush Administration endorsed the concept of "no-net-loss" of wetlands— a goal declared by President George H. W. Bush in 1988 and also embraced by President Clinton to balance wetlands losses and gains in the short term and achieve net gains in the long term. On Earth Day 2004, President Bush announced a new national goal, moving beyond no-net-loss to achieve an overall increase of wetlands. The goal was to create, improve, and protect at least 3 million wetland acres over the next five years in order to increase overall wetland acres and quality. (By comparison, the Clinton Administration in 1998 announced policies intended to achieve overall wetland increases of 200,000 acres per year by 2005.) The Bush strategy also called for better tracking of wetland programs and enhanced local and private sector collaboration.

In April 2008, the Bush Administration issued a report saying that more than 3.6 million acres of wetlands had been restored, protected, or improved as part of the President's program to create, improve and protect wetlands, and that the number was expected to climb to 4.5 million acres by the original date set by that program—Earth Day 2009.[6] The report documented gains, but not offsetting loses. It summarized accomplishments for each federal wetland conservation program. Environmental groups criticized the report as presenting an incomplete picture, because it failed to mention wetlands lost to agriculture and development.

Numerous shifts in federal policies since 1985 (and changes in economic conditions as well) strongly influence wetland loss patterns, but the composite effects remain unmeasured beyond these raw numbers. There usually is a large time lag between the announcement and implementation of changes in policy, and collection and release of data that measure how these changes affect loss rates. Also, it is often very difficult to distinguish the role that policy changes play from other factors, such as agricultural markets, development pressures, and land markets.

Further, these data only measure acres. This may have been appropriate two or three decades ago when scientists knew less about how to measure the specific functions and values found in wetlands. By providing data limited to number of acres, these data provide few insights into changes in their quality, as measured by the values they provide, which is often determined by factors such as where a wetland is located in a watershed, and what are the surrounding land uses. Scientists caution that there are a number of questions about the qualitative and ecological integrity of existing wetlands. The wetlands trends data reported by FWS in 2011 show increases in certain types of freshwater wetlands since 2004, particularly freshwater ponds constructed to replace lost wetlands. However, FWS noted that there is no clear scientific consensus about the functional equivalency of replacement wetlands.

Wetlands and Climate Change

As described above, coastal wetlands provide critical services such as absorbing energy from coastal storms, preserving shorelines, protecting human populations and infrastructure, absorbing pollutants, and serving as critical habitat for migratory species. Many scientists believe that these resources and services will be threatened as sea-level rise associated with a changing climate inundates wetlands. Due in part to their limited capacity for adaptation, wetlands have been considered among the ecosystems most vulnerable to a changing climate. Changes in climatic conditions that affect water conditions (e.g., wetter, drier, more saline) are predicted to have substantial impact on species that use wetlands and on ecosystem services provided by wetlands, or make efforts to reestablish wetlands more challenging.[7]

In 2010, a group of international scientists published results of research modeling efforts to identify conditions under which coastal wetlands could survive rising sea level. Using a rapid sea-level rise scenario, the scientists estimated that most coastal wetlands worldwide will experience inundation that leads to rapid and irreversible conversion of marshland into unvegetated, subtidal surfaces and will disappear near the end of the 21st century. Under moderate and slow sea-level rise scenarios, some coastal wetlands would be vulnerable to inundation, depending on amounts of sediment present: larger amount of sediment would enable the wetland to adapt and modify naturally and thus be more likely to survive sea-level rise.[8]

Coastal wetlands also serve as a "sink" for absorbing carbon dioxide (CO_2), the most common greenhouse gas (GHG) that is associated with climate change. Scientists recognize that tidal wetlands hold large amounts of carbon, some within standing plant biomass, but most within deep organic-bearing soils. Carbon that is stored in soils has been built up over millennia and reflects pools of CO_2 that have been transferred from the atmosphere and sequestered within roots and other organic material.[9] However, the loss of wetland areas, for example through inundation and erosion, eliminates its ongoing sequestration capacity, and draining wetlands for development releases within a few decades carbon that took centuries to accumulate. In 2011, the World Bank released a report which concluded that drainage and degradation of coastal wetlands has become a major cause of carbon dioxide emissions that contribute to climate change, large enough globally that carbon dioxide emissions from drained coastal wetlands should be included in carbon accounting and emission inventories, and in policy frameworks to reduce emissions.[10] Some policymakers concerned with mitigating climate change have begun to consider whether it is possible to halt the release of carbon from converted or eroded wetlands and reverse carbon losses through wetland restoration. Further, some are considering whether the ecosystem benefits of wetlands, from a carbon sequestration standpoint, can be quantified in financial terms to enable use of wetlands restoration and management as potential generators of GHG offsets in the context of climate change policy.[11]

SELECTED FEDERAL WETLANDS PROGRAMS

Federal program issues include the administration of programs to protect, restore, or mitigate wetland resources (especially the Clean Water Act Section 404 program); relationships between agricultural and regulatory programs; whether all wetlands should be treated the same in federal programs, and which wetlands should be subject to regulation; and whether protecting wetlands by acres is an effective proxy for protecting wetlands based on the functions they perform and the values they provide. In addition, private property questions are raised, because almost three-quarters of the remaining wetlands are located on private lands. Some property owners believe that they should be compensated when federal programs limit how they can use their land and for decisions that arguably diminish the value of the land.

The Clean Water Act Section 404 Program

The principal federal program that provides regulatory protection for wetlands is found in Section 404 of the Clean Water Act (CWA). Its intent is to protect water and adjacent wetland areas from adverse environmental effects due to discharges of dredged or fill material. Enacted in 1972, Section 404 requires landowners or developers to obtain permits from the Corps of Engineers to carry out activities involving disposal of dredged or fill materials into waters of the United States, including wetlands.

The Corps has long had regulatory jurisdiction over dredging and filling, starting with the River and Harbor Act of 1899. The Corps and EPA share responsibility for administering the Section 404 program. Other federal agencies, including NRCS, FWS, and NMFS, also have roles in this process. In the 1970s, legal decisions in key cases led the Corps to revise this program to incorporate broad jurisdictional definitions in terms of both regulated waters and adjacent wetlands. Section 404 was last amended in 1977.

This judicial/regulatory/administrative evolution of the Section 404 program has generally pleased those who view it as a critical tool in wetland protection, but dismayed others who would prefer more limited Corps jurisdiction or who see the expanded regulatory program as intruding on private land-use decisions and treating wetlands of widely varying value similarly. Underlying this debate is the more general question of whether Section 404 is the best approach to federal wetland protection.

Some wetland protection advocates have proposed that it be replaced or greatly altered. First, they point out that it governs only the discharge of dredged or fill material, while not regulating other acts that drain, flood, or otherwise reduce functional values. Second, because of exemptions provided in 1977 amendments to Section 404, major categories of activities are not required to obtain permits. These include normal, ongoing farming, ranching, and silvicultural (forestry) activities. Further, permits generally are not required for activities that drain wetlands (only for those that fill wetlands), which excludes a large number of actions with potential to alter wetlands. Third, in the view of protection advocates, the multiple values that wetlands can provide (e.g., fish and wildlife habitat, flood control) are not effectively recognized through a statutory approach based principally on water quality, despite the broad objectives of the Clean Water Act.

The Permitting Process

The Corps' regulatory process involves both general permits for actions by private landowners that are similar in nature and will likely have a minor effect on wetlands, and individual permits for more significant actions. According to the Corps, it evaluates more than 85,000 permit requests annually. Of those, more than 90% are authorized under a general permit, which can apply regionally or nationwide, and is essentially a permit by rule, meaning the proposed activity is presumed to have a minor impact, individually and cumulatively. Most general permits do not require pre-notification or prior approval by the Corps. About 9% of all permits are required to go through the more detailed evaluation for a standard individual permit, which may involve complex proposals or sensitive environmental issues and can take 180 days or longer for a decision. Less than 0.3% of permits are denied; most other individual permits are modified or conditioned before issuance. About 5% of applications are withdrawn prior to a permit decision.

Regulatory procedures on individual permits allow for interagency review and comment, a coordination process that can generate delays and an uncertain outcome, especially for environmentally controversial projects. EPA is the only federal agency having veto power over a proposed Corps permit; EPA has used its veto authority 13 times in the 40-plus years since the program began. However, critics have charged that implied threats of delay by the FWS and others practically amount to the same thing. Reforms during the Reagan, George H. W. Bush, and Clinton Administrations streamlined certain of these procedures, with the intent of speeding up and clarifying the Corps' full regulatory program, but concerns continue over both process and program goals.

Controversy also surrounded revised regulations issued by EPA and the Corps in 2002, which redefine two key terms in the 404 program: "fill material" and "discharge of fill material." These definitions are important, because material defined as "fill" is regulated and permitted under Section 404 procedures, while other waste discharges are regulated under more stringent CWA rules and procedures. The agencies said that the revisions were intended to clarify certain confusion in their joint administration of the program due to previous differences in how the two agencies defined those terms. However, environmental groups contended that the changes allow for less restrictive and inadequate regulation of certain disposal activities, including disposal of coal mining waste, which could be harmful to aquatic life in streams. Legislation to reverse the agencies' action by clarifying in the law that fill material cannot be composed of waste has been introduced regularly since the 107th Congress, including H.R. 1837 in the 113th Congress.[12]

As previously described, three criteria—hydrology, soil type, and plants—are used in making wetlands delineations under several environmental laws and programs, including Section 404 permitting. Scientists generally agree that each of the three parameters must be met to identify an area as a wetland. Because growth of plants in wetland areas typically is contingent on the presence of hydric soils and the availability of sufficient water, the vegetation parameter often is determinative of whether an area qualifies as a wetland or not.

In 2012, the Corps revised the National Wetlands Plant List (NWPL), which is used by federal and state agencies for determining whether a particular area contains a prevalence of hydrophytic (i.e., wetland) vegetation. This is the first major revision of the plant list since its publication in 1988 and is intended to improve the accuracy of the overall list. The updated list contains 8,200 plant species, an increase of 1,472 species, or 22%, primarily as a result of new taxonomic interpretations. The Corps said it does not expect major changes to wetland delineations as a result of the updated list, but some commenters contend that the new list is likely to cause more areas to qualify as wetlands.[13]

Section 404 authorizes states to assume many of the Corps' permitting responsibilities. Two states have done this: Michigan (in 1984) and New Jersey (in 1992). Others reasons cited for not joining these two states include the complex process of assumption, the anticipated cost of running a program, and the continued involvement of federal agencies because of statutory limits on waters that states could regulate. Efforts continue to encourage more states to assume program responsibility.

Nationwide Permits

Nationwide permits are a key means by which the Corps minimizes the burden of its regulatory program. A nationwide permit is a form of general permit that authorizes a category of activities throughout the nation and is valid only if the conditions applicable to the permit are met. They are issued for periods of no longer than five years and thereafter must be reissued by the Corps. These general permits authorize activities that are similar in nature and are judged to cause only minimal adverse effect on the environment, individually and cumulatively. General permits authorize landowners to proceed without having to obtain individual permits in advance.[14]

The current nationwide permit program has few strong supporters, for differing reasons. Developers say that it is too complex and burdened with arbitrary restrictions. Environmentalists say that it does not adequately protect aquatic resources. At issue is whether the program has become so complex and expansive that it cannot either protect aquatic resources or provide for a fair regulatory system, which are its dual objectives.

Section 404 Judicial Proceedings: SWANCC and *Rapanos*

The Section 404 program has been the focus of numerous lawsuits, most of which have sought to narrow the geographic scope of the regulatory program.

SWANCC

An issue of long-standing controversy is whether isolated waters are properly within the jurisdiction of Section 404. Isolated waters (those that lack a permanent surface outlet to downstream waters) that are not physically adjacent to navigable surface waters often appear to provide few of the values for which wetlands are protected, even if they meet the technical definition of a wetland. In January 2001, the Supreme Court ruled on the question of whether the CWA provides the Corps and EPA with authority over isolated waters and wetlands. The Court's 5-4 ruling in *Solid Waste Agency of Northern Cook County (SWANCC) vs. U.S. Army Corps of Engineers* (531 U.S. 159 (2001)) held that the denial of a Section 404 permit for disposal on isolated wetlands solely on the basis that migratory birds use the site exceeds the authority provided in the CWA. The full extent of retraction of the regulatory program resulting from this decision remains unclear, even more than a dozen years after the ruling. Environmentalists believe that the Court misinterpreted congressional intent on the matter, while industry and landowner groups welcomed the ruling.[15]

Policy implications of how much the decision restricts federal regulation depend on how broadly or narrowly the opinion is applied, and, since the 2001 Court decision, other federal courts have issued a number of rulings that have reached varying conclusions. Some federal courts have interpreted *SWANCC* narrowly, thus limiting its effect on current permit rules, while a few read the decision more broadly. Attorneys for industry and developers say that the courts will remain the primary battleground for CWA jurisdiction questions, so long as neither the Administration nor Congress takes steps to define jurisdiction.

The government's view on the key question of the scope of CWA jurisdiction in light of *SWANCC* and other court rulings came in a legal memorandum issued jointly by EPA and the Corps in January 2003.[16] It provided a legal interpretation essentially based on a narrow reading of the Court's decision, thus allowing federal regulation of some isolated waters to continue (in cases where factors other than the presence of migratory birds may exist, thus allowing for assertion of federal jurisdiction), but it called for more review by higher levels in the agencies in such cases. Administration press releases said that the guidance demonstrates the government's commitment to "no-net-loss" wetlands policy. However, it was apparent that the issues remained under discussion, because at the same time, the Administration issued an advance notice of proposed rulemaking (ANPRM) seeking comment on how to define waters that are under the regulatory program's jurisdiction. The ANPRM did not actually propose rule changes, but it indicated possible ways that Clean Water Act rules might be modified to further limit federal jurisdiction, building on *SWANCC* and some of the subsequent legal decisions. The government received more than 133,000 comments on the ANPRM, most of them negative, according to EPA and the Corps. Environmentalists and many states opposed changing any rules, saying that the law and previous court rulings call for the broadest possible interpretation of the Clean Water Act (and narrow interpretation of *SWANCC*), but developers sought changes to clarify interpretation of the *SWANCC* ruling.

In December 2003, EPA and the Corps announced that the Administration would not pursue rule changes concerning federal regulatory jurisdiction over isolated wetlands. The EPA Administrator said that the Administration wanted to avoid a contentious and lengthy rulemaking debate over the issue. Nonetheless, interest groups on all sides have been critical of confusion in implementing the 2003 guidance, which constitutes the main tool for interpreting the reach of the *SWANCC* decision. Environmentalists remain concerned about diminished protection resulting from the guidance, while developers said that without a new rule, confusing and contradictory interpretations of wetland rules likely will continue. In that vein, a Government Accountability Office (GAO) report concluded that Corps districts differ in how they interpret and apply federal rules when determining which waters and wetlands are subject to federal jurisdiction, documenting enough differences that the Corps undertook a comprehensive survey of its district office practices to help promote greater consistency.[17] Concerns over inconsistent or confusing regulation of wetlands have also drawn congressional interest.[18]

Rapanos-Carabell

Federal courts continue to have a key role in interpreting and clarifying the *SWANCC* decision. In February 2006, the Supreme Court heard arguments in two cases brought by landowners (*Rapanos vs. United States*; *Carabell vs. U.S. Army Corps of Engineers*) seeking to narrow the scope of the CWA permit program as it applies to development of wetlands. The issue in both cases had to do with the reach of the CWA to cover "waters" that were not navigable waters, in the traditional sense, but were connected somehow to navigable waters or "adjacent" to those waters. (The act requires a federal permit to discharge dredged or fill materials into "navigable waters.") Many legal and other observers hoped that the Court's ruling in these cases would bring greater clarity about the scope of federal regulatory jurisdiction.

The Court's ruling was issued on June 19, 2006 (*Rapanos et al., vs. United States*, 547 U.S. 715 (2006)). In a 5-4 decision, a plurality of the Court, led by Justice Scalia, held that the lower court had applied an incorrect standard to determine whether the wetlands at issue are covered by the CWA. Justice Kennedy joined this plurality to vacate the lower court decisions and remand the cases for further consideration, but he took different positions on most of the substantive issues raised by the cases, as did four other dissenting justices.[19] Legal observers suggested that the implications of the ruling (both short-term and long-term) are far from clear. Because the several opinions written by the justices did not draw a clear line regarding what wetlands and other waters are subject to federal jurisdiction, one result has been more case-by-case determinations and continuing litigation.

In December 2008, EPA and the Corps issued guidance to enable their field staffs to make CWA jurisdictional determinations in light of the decision. According to the nonbinding guidance, the agencies would assert regulatory jurisdiction over certain waters, such as traditional navigable waters and adjacent wetlands. Jurisdiction over others, such as non-navigable tributaries that do not typically flow year-round and wetlands adjacent to such tributaries, would be determined on a case-by-case basis, to determine if the waters in question have a significant nexus with a traditional navigable water. The guidance details how the agencies should evaluate whether there is a significant nexus. The guidance was not intended to increase or decrease CWA jurisdiction, and it did not supersede or nullify the January 2003 guidance memorandum, discussed above, which addressed jurisdiction over isolated wetlands in light of *SWANCC*. The guidance takes the view that waters are jurisdictional if they satisfy *either* the plurality or Kennedy tests in *Rapanos*. The 2008 guidance also provides detail for determining whether a wetland is adjacent to a traditional navigable water and whether a tributary of a navigable water is subject to the act—key issues raised by the *Rapanos* decision.

In April 2011, the Obama Administration weighed into the CWA jurisdiction debate as EPA and the Corps proposed new joint agency guidance to clarify regulatory jurisdiction over U.S. waters and wetlands and to replace the agencies' 2008 guidance. Like the existing guidance, the proposed revisions would adopt the Kennedy-test-or-plurality-test view of interpreting *Rapanos*. However, the agencies believed that a wider evaluation of jurisdiction is possible than the existing guidance suggests, stating, "after careful review of these opinions, the agencies concluded that previous guidance did not make full use of the authority provided by the CWA to include waters within the scope of the Act, as interpreted by the Court."[20]

EPA and the Corps acknowledged that, compared with the existing guidance, the proposed guidance revisions were likely to increase the number of waters identified as protected by the CWA—a conclusion that pleased some observers but alarmed others. Supporters believed that new guidance will improve protection of U.S. waters and wetlands, while critics argue that it represents overreaching, beyond authority provided by Congress. Others faulted the continued reliance on federal guidance, which is not binding and lacks the force of law, yet can have significant impact on regulated entities. Final guidance was submitted to the White House Office of Management and Budget (OMB) for review in February 2012, but it was not released. Instead, in September 2013, EPA and the Corps withdrew the revised guidance document from OMB review in favor of regulatory revisions, which the agencies proposed in March 2014.

SWANCC and *Rapanos* generated confusion beyond what already existed as to the reach of "waters of the United States." The lack of a majority rationale in *Rapanos* has led lower courts to extract different tests from the decision for measuring this reach, and Justice Kennedy's "significant nexus" concept remains amorphous and undefined. The EPA-Corps 2011 draft guidance—now withdrawn—was intended to reduce the confusion, but many observers and stakeholders contend that jurisdictional issues remain in dispute throughout the country, leading to costly project delays and uncertain protection of wetland resources.

The 2014 proposed rule would replace the existing guidance, which remains in effect. According to the agencies, the proposed rule would revise the existing administrative definition of "waters of the United States" consistent with the Supreme Court's rulings and consistent with science concerning the interconnectedness of tributaries, wetlands, and other waters and the effects of these connections on the chemical, physical, and biological integrity of downstream waters. The agencies believe that the proposed rule does not protect any new types of waters that have not been protected historically and that it does not exceed the CWA's coverage. While it would enlarge categorical jurisdiction beyond that under the 2003 and 2008 EPA-Corps guidance, which the agencies believe was narrower than is justified by science and the law, they also believe that it would not enlarge jurisdiction beyond what is consistent with the Supreme Court's narrow reading of jurisdiction. Others disagree. A number of agriculture and other industry groups contend that the proposed rule would expand CWA jurisdiction and that many terms and definitions in the rule are unclear, causing concern and uncertainty. The agencies are accepting public comment on the rule until November 14, 2014 and could issue a final rule in 2015.[21]

While the issue of how regulatory protection of wetlands is affected by the *SWANCC* and *Rapanos* decisions continues to evolve, the remaining responsibility to protect affected wetlands falls on states and localities. Whether states will act to fill in the gap left by removal of some federal jurisdiction is likely to be constrained by budgetary and political pressures, but after *SWANCC* a few states (Wisconsin and Ohio, for example) passed new laws or amended regulations to do so. Many states have said that they do not have authority or financial resources to protect their wetlands, in the absence of federal involvement.

Congressional Response

Legislation to reverse the *SWANCC* and *Rapanos* decisions was introduced on several occasions since the 107th Congress. In the 111th Congress, the Senate Environment and Public Works Committee approved S. 787—the first such proposal to advance from a congressional committee. Companion House legislation was introduced in the 111th Congress (H.R. 5088). There was no further legislative action on either bill.[22]

In 2009, Administration officials urged Congress to consider four general principles in legislation to clarify waters protected by the CWA:

- Broadly protect the nation's waters;
- Make the definition of covered waters predictable and manageable;
- Promote consistency between CWA and agricultural wetlands programs; and
- Recognize long-standing practices, such as exemptions now in effect through regulations and guidance.[23]

In light of the widely differing views of proponents and opponents, future prospects for legislation on the geographic scope of CWA jurisdiction are highly uncertain. Critics questioned the constitutionality of legislation that was proposed and asserted that it would expand federal authority, thus likely increasing confusion, rather than settling it. An additional difficulty of legislating changes to the CWA in order to protect wetlands results from the fact that the complex scientific questions about such areas (see discussion above, "Wetlands: Science and Information") are not easily amenable to precise resolution in law. The debate over revising the act highlights the challenges of using the law to try to do so.

Recent efforts by EPA and the Corps to develop revised *Rapanos* guidance (which has now been withdrawn) have been controversial in Congress and elsewhere. Legislative provisions to prohibit the agencies from funding activities related to revising the guidance were included in several appropriations bills in the 112th Congress, but none of these provisions was enacted. Interest in similar legislation concerning the guidance has continued with bills in the 113th Congress, such as S. 1006 and H.R. 1829, to prevent the agencies from finalizing the now-withdrawn 2011 draft guidance, and S. 890/H.R. 3377, which would amend the CWA with a narrow definition of waters that are subject to the act's jurisdiction. Legislation addressing the EPA-Corps proposed rule also has been introduced. H.R. 5078 would prohibit the agencies from finalizing the proposed "waters" rule or using that proposal or similar proposed rule or guidance as the basis of any rulemaking on the scope of CWA jurisdiction, and it would require withdrawal of a related interpretive rule that applies to agriculture. It also would direct the Corps and EPA to consult with state and local officials on CWA jurisdiction issues and develop a report on results of such consultation. By a 262-152 vote, the House passed this bill on September 9.

Should All Wetlands Be Treated Equally?

Under the Section 404 program, there is a perception that all jurisdictional wetlands are treated equally, regardless of size, functions, or values. In reality, this is not the case, because the Corps' general permits do provide accelerated regulatory decisions for many activities that affect wetlands. Further, a number of types of activities are fully exempt from 404 permit requirements as a result of statutory provisions enacted in 1977 (including ongoing farming, ranching, and forestry activities, as specified in Section 404(f)) and regulatory exemptions (including for prior converted croplands, which are wetlands that were drained, dredged, filled, leveled, or otherwise manipulated before

December 23, 1985, to make production of an agricultural commodity possible).

However, this perception has led critics to focus on situations where a wetland has little apparent value, but the landowner's development proposal is not approved, or the landowner is penalized for altering a wetland without a federal permit. Critics believe that one possible solution may be to have a tiered approach for regulating wetlands. Legislation introduced in past Congresses proposed to establish multiple tiers (typically three)—from highly valuable wetlands that should receive the greatest protection to the least valuable wetlands where alterations might usually be allowed. Some states (New York, for example) use such an approach for state-regulated wetlands.

Three questions arise: (1) What are the implications of implementing a classification program? (2) How clearly can a line separating each wetland category be defined? (3) Are there regions where wetlands should be treated differently? Regarding classification, even many wetland protection advocates acknowledge that there are some situations where a wetland designation with total protection is not appropriate. But they fear that classification for different degrees of protection could be a first step toward a major erosion in overall wetland protection. Also, these advocates would probably like to see almost all wetlands presumed to be in the highest protection category unless experts can prove an area should receive a lesser level of protection, while critics who view protection efforts as excessive would seek the reverse. In response to these concerns, Corps and EPA officials note that existing guidance and regulations already provide substantial flexibility to implement current programs, allowing, for example, less vigorous permit review to small projects with minor environmental impacts. Some types of wetlands are already treated differently—for example, playas and prairie potholes, which have somewhat different definitions under swampbuster (discussed below). However, this differential treatment contributes to questions about federal regulatory consistency on private property.

Locating the boundary line of a wetland can be controversial when the line encompasses areas that do not meet the image held by many. Controversy would likely grow if a tiered approach required that lines segment wetland areas. On the other hand, a consistent application of an agreed-on definition might lead to fewer disputes and result in more timely decisions.

Some states have far more wetlands than others. Different treatment has been proposed for Alaska, because about one-third of the state is designated as wetlands, yet a very small portion has been converted. In the past, legislative proposals have been made to exempt that state from the Section 404 program until 1% of its wetlands have been lost.

Agriculture and Wetlands

National surveys more than two decades ago indicated that agricultural activities had been responsible for about 80% of wetland loss in the preceding decades, making this topic a focus for policy makers seeking to protect the remaining wetlands. Congress responded by creating wetland conservation programs in farm legislation starting in 1985. Conservation programs in the farm bill use both incentives and disincentives to encourage landowners to protect and restore wetlands. For example, the Conservation Reserve Program's wetland program uses incentives to protect wetlands, while swampbuster uses disincentives.[24]

Members of the farm community have expressed a wide range of views about wetland protection, from strong opposition to strong support. These views are frequently framed in the context of two general concerns about wetland protection efforts. First, as a philosophical matter, some object to federal regulation of private lands, regardless of the societal values those lands might provide. Second, many farmers want certainty and predictability about the land they farm to limit their financial risk. Therefore, if wetlands are located on farm property, they want assurances that the boundary line delineating wetlands will remain where located for as long as possible.

Swampbuster

Swampbuster, enacted in 1985, uses disincentives rather than regulations to protect wetlands on agricultural lands. It removes a farmer's eligibility from all government price and income support programs for activities such as draining, dredging, filling, leveling or otherwise altering a wetland. Producers who plant a program crop on a wetland converted to agricultural use after December 23, 1985, or who convert wetlands, making agricultural commodity production possible, after November 28, 1990, are ineligible for certain U.S. Department of Agriculture (USDA) program benefits. Swampbuster has been controversial with farmers concerned about redefining an appropriate federal role in wetland protection on agricultural lands, and with wetland protection advocates concerned about inadequate enforcement. Since 1995, the NRCS has made wetland determinations only in response to requests because of uncertainty over whether changes in regulation or law would modify boundaries that have already been delineated.

Swampbuster amendments in 1996 (P.L. 104-127) granted producers greater flexibility by making changes, such as exempting swampbuster penalties when wetlands are voluntarily restored; providing that prior converted wetlands are not to be considered "abandoned" if they remain in agricultural use; and granting good-faith exemptions. They also encourage mitigation, established a mitigation banking pilot program, and repealed required consultation with the FWS.

The 113th Congress enacted legislation to renew the farm bill (P.L. 113-79), including limited modifications to the swampbuster program. The enacted bill adds crop insurance premium subsides as an ineligible benefit, if found to be out of compliance. The wetlands compliance provision of the farm bill includes a number of exempt lands, such as wetlands created by irrigation delivery systems. The 2014 farm bill amendments extend the list of exemptions for compliance violators, allowing additional time for producers to remedy or mitigate the wetland conversion before losing crop insurance premium subsidies.[25]

Other Agricultural Wetlands Programs

Several USDA conservation programs provide federal payments to private agricultural landowners for voluntary changes in land use or management to achieve environmental benefits, including wetlands protection.[26] The Wetland Reserve Program (WRP), enacted in 1990, provided landowners with payments for placing easements on farmed wetlands. It provided long-term technical and financial assistance to landowners with the opportunity to protect, restore, and enhance wetlands on their property, and to establish wildlife practices and protection. Strong farmer interest led Congress to raise the WRP enrollment ceiling in both the 2002 and 2008 farm bills. The 2008 legislation authorized a Wetlands Reserve Enhancement Program, allowing USDA to enter into agreements with states in order to leverage federal funds for wetlands protection and enhancement.

Farm bill legislation in the 113th Congress (P.L. 113-79) modified agriculture conservation programs in several respects. The legislation repealed the WRP and two other easement programs and created a new Agricultural Conservation Easement Program (ACEP). The new program retains wetlands reserve easements similar to WRP to protect and restore wetlands, along with agricultural land easements for the other repealed programs. Program participants agree to restore and maintain wetlands according to an approved wetland reserve easement plan, while in return, USDA provides technical and finance assistance for wetland restoration. Landowners are compensated for the wetland reserve easement based on the fair market value of the land and the length of the easement or contract.

The farm bill also includes the Environmental Quality Incentives Program (EQIP), which may have incidental protection benefits for wetlands. For example, EQIP supports the installation or implementation of structural and management practices, and the 2008 farm bill expanded the program to include practices that enhance wetlands. The 2014 farm bill reauthorized the financial and technical assistance elements of EQIP and also incorporated into EQIP a previously separate program, the Wildlife Habitat Incentives Program (WHIP), which provides assistance to landowners for development of wetland wildlife and other types of wildlife habitat. As amended, the farm bill now requires that 5% of total EQIP payments benefit wildlife habitat. Finally, some programs could less directly help protect wetlands, such as the Conservation Stewardship Program, which provides payments to maintain and improve existing conservation practices on agricultural lands and to adopt additional conservation activities.

Building on programs expressly authorized in the farm bill, USDA administers several conservation programs that it established administratively. One is the Wetland Restoration, NonFloodplain Initiative to allow enrollment of up to 250,000 acres of large wetland complexes and playa lakes located outside the 100-year floodplain in the CRP. CRP, the largest federal, private-land retirement program in the United States, allows producers to enter into 10- to 15-year contracts to install certain conservation practices. As of July 2014, there were 245,000 acres enrolled in the Wetland Restoration, Non-Floodplain Initiative. USDA also established a Wetland Restoration Initiative to enroll wetlands located in the 100-year floodplain in the CRP. As of July 2014, a total of 265,000 acres were enrolled. Participants in these programs receive incentive payments equal to 25% of the cost to help pay for restoring the hydrology of the site, as well as rental payments and cost-sharing assistance to install eligible conservation practices.

Agricultural Wetlands and the Section 404 Program

The CWA Section 404 program applies to qualified wetlands in all locations, including agricultural lands. But the Corps and EPA exempt "prior converted lands" (wetlands modified for agricultural purposes before 1985) from Section 404 permit requirements under a memorandum of agreement (MOA) by rule, and since 1977 the CWA has exempted "normal farming activities" from Section 404. The Supreme Court's *SWANCC* decision exempted certain isolated wetlands from Corps jurisdiction; NRCS estimated that about 8 million acres in agricultural locations might be exempted by this decision.

While these exemptions and the MOA displease some protection advocates, they probably dampened some of the criticism from farming interests over federal regulation of private lands, at least for a while. On the other hand, the prospect that Congress might enact legislation to reverse the Court's 2001 and 2006 rulings, discussed above, particularly alarmed farm groups, who fear that changes in law or regulations could negatively affect their activities. Because of differences between the CWA and farm bill on the jurisdictional status of certain wetlands (e.g., isolated wetlands may be regulated differently by federal agencies), in 2005 the Corps and NRCS signed a memorandum of understanding and issued joint guidance clarifying circumstances where wetlands delineation made by one agency can be accepted for determining the jurisdiction of the other agency.[27]

Recently, agriculture industry groups have been among the most vocal critics of the EPA-Army Corps proposed "waters of the United States" rule, discussed above, out of concern that the proposal would bring agricultural lands and activities under Clean Water Act jurisdiction.

Other Federal Protection Efforts

Many federal agencies have been active in wetland improvement efforts in recent years. In particular, the Fish and Wildlife Service (FWS) has been promoting the success of its Partners for Fish and Wildlife program, which Congress reauthorized through FY2011 in 2006 (P.L. 109-294). Through voluntary agreements, the Partners program provides technical assistance and cost-share incentives directly to landowners for wetland restoration projects on private lands.[28]

FWS also administers the National Coastal Wetlands Conservation Grant Program, established by Title III of P.L. 101-646, Coastal Wetlands Planning, Protection and Restoration Act of 1990. Under this program, federal grants, matched by contributions from states and localities, as well as from private landowners and conservation groups, are used to acquire, restore, or enhance coastal wetlands and adjacent uplands to provide long-term conservation benefits to fish, wildlife, and their habitats. The federal government generally provides 50% of the total costs of a project, but the federal share can be increased to 75% if the state maintains a fund for acquiring coastal wetlands. Since 1992, about $183 million in grants have been awarded to 25 coastal states and one U.S. territory for projects involving 250,000 acres of coastal wetland ecosystems.[29]

Other programs also restore and protect domestic and international wetlands. One of these derives from the North American Wetlands Conservation Act, reauthorized through FY2012 in P.L. 109- 322 with an appropriations ceiling of $75 million annually. This act provides matching grants to organizations and individuals who have developed partnerships to carry out wetlands conservation projects in the United States, Canada, and Mexico for the benefit of wetlands-associated migratory birds and other wildlife. Grants consist of standard grants, which support projects in all three countries, and small grants, which support similar activities but at a smaller scale and for fewer dollars. According to the FWS, from September 1990 through March 2014, approximately 5,000 partners in 2,421 projects have received nearly $1.3 billion in grants. They have contributed another $2.7 billion in matching funds to affect 27.5 million acres of habitat. [30]

Under the Convention on Wetlands of International Importance, more commonly known as the Ramsar Convention, the United States is one of 168 nations that have agreed to slow the rate of wetlands loss by designating wetland sites of international importance. These nations have designated 2,186 sites, totaling 515 million acres, since the convention was adopted in 1971. The United States has designated 36 sites pursuant to the convention, encompassing 4.5 million acres.

Private Property Rights and Landowner Compensation

An estimated 74% of all remaining wetlands in the conterminous states are on private lands. Questions of federal regulation of private property stem from the argument that landowners should be compensated when a "taking" occurs and alternative uses are prohibited or restrictions on use are imposed to protect wetland values. The U.S. Constitution provides that property owners shall be compensated if private property is "taken" by government action. The courts generally have found that compensation is not required unless all reasonable uses are precluded. Many individuals or companies purchase land with the expectation that they can alter it. If that ability is denied, they contend, then the land is greatly reduced in value. Many argue that a taking should be recognized when a site is designated as a wetland.

Congress has explored these wetlands property rights issues on several occasions. An example is a 2001 hearing by the House Transportation and Infrastructure Committee, Subcommittee on Water Resources and the Environment.[31] Recent Congresses considered, but did not enact, property rights protection proposals.

STATE PROTECTION EFFORTS

In addition to federal programs and activities, wetlands in the United States are regulated and protected through a variety of state and local laws and regulations, as well as through initiatives and actions of nongovernmental organizations, schools and universities, and private citizens. The role of states in wetland protection is especially important, as noted in a study by the Environmental Law Institute.

> States have long held the right and the responsibility to provide stewardship over their resources, and state agency staff typically have a well-versed understanding of the "lay of the land," in terms of both topography and state priorities, policies, and practices. Finally, in light of recent uncertainty over federal jurisdiction of wetlands and limited federal resources for wetland protection, the role of states in conserving wetlands may be more important now than ever before.[32]

States use a variety of programs and tools to protect and manage wetlands, including regulation and mitigation, wetland water quality standards, monitoring and assessment, voluntary restoration, tax incentives, coordination among state and federal agencies, and public/private partnerships. Programs vary substantially from state to state and often derive their authorities from more than one statute and/or regulation. As a result, different programs may be administered by different state agencies. In addition, programs may change from year to year.[33]

Every state regulates, to some degree, activities that affect wetlands, but two-thirds of the states lack regulatory programs that *comprehensively* regulate wetlands. Many states rely solely or primarily on authority in CWA Section 401, under which states may review any activity that requires a federal permit or license to determine its effect on the state's water quality standards.[34] Section 401 gives states the authority to approve, condition, or deny the federal permit— including a Section 404 permit—or license based on their review. In areas where there is no Section 404 permit requirement, and therefore no opportunity for review under Section 401, some states also require a state permit for activities that affect aquatic resources: 23 states have authority to issue permits for dredge and fill activities in wetlands and other waters of the state, such as geographically isolated wetlands (although as described previously, only New Jersey and Michigan have been delegated 404 permitting authority).

As is the case with the federal regulatory program under CWA Section 404, an important consideration is how a state determines which waters fall within its regulatory jurisdiction. States' definitions of their waters are typically much broader than the federal definition of "waters of the United States," meaning that states may exert jurisdiction over waters within their boundaries that are not covered by the CWA. State definitions often include phrases such as "all surface waters." They also may exclude certain waters, such as private lakes or ponds. Groundwater is not included in the federal regulatory definition, but most states include groundwater in their regulatory programs.[35] All 50 states include wetlands in either or both their statutory and regulatory definitions of state waters—32 make this inclusion explicit, and 18 define waters more generally, including wetlands implicitly. The inclusion of wetlands in a state's definition of state waters does not give automatic protection to these waters; the state must also have some form of complementary regulatory authority, such as to issue permits.[36]

Other findings of the ELI report include the following.

- The majority of states have adopted legislation, policies, and/or guidelines for mitigating impacts to aquatic resources that are permitted in their states. Mitigation provisions range from general requirements to specific replacement ratios, site preferences, and mitigation options such as purchasing credits from a mitigation bank (also see "Wetland Restoration and Mitigation").
- One-third of states report having a wetland-specific monitoring and/or assessment program or monitoring wetlands as part of a larger state monitoring program.
- Nearly one-half of the states operate a formal program for partnering with private landowners on restoration or conservation, and a majority of states report that they conduct outreach or provide technical assistance to private landowners. Ninety percent of states have one or more agencies that carry out education and outreach activities related to wetlands.

The Louisiana Experience

Much of the attention to reversing wetland loss has focused on Louisiana, where an estimated 80% of the total loss of U.S. coastal wetlands has occurred and where about 40% of U.S. coastal wetlands that remain in the lower 48 states are located (coastal wetlands are about 5% of all U.S. wetlands). Changes to Louisiana's coastal area result from a combination of natural environmental processes (erosion, saltwater intrusion into fresh systems, sea level rise) and human-related activities, according to the U.S. Geological Survey (USGS). Wetland loss has occurred naturally for centuries, but until recently, land losses have been counterbalanced by various natural wetland-building processes.

USGS estimates that, since 1932, coastal Louisiana has experienced a net change in land area of approximately 1,883 square miles—an area the size of Delaware. Land loss rates on the Louisiana coast have slowed from an average of more than 30 square miles per year between 1956 and 1978, to an estimated 11.8 square miles per year from 1985 to 2004. When the hurricanes of 2005 and 2008 are factored in, the trend increased the amount of land loss to 16.6 square miles from 1985 to 2010. According to USGS, if this loss were to occur at a constant rate, it would equal losing more than a football field every hour.[37] As a result of wetlands loss, the natural flow of Mississippi River and floodwaters to feed sediment to the marshes has been reduced. Saltwater has invaded the brackish estuaries, destroying vegetation and areas that are needed for fish, shellfish, and wildlife. In response to these losses, Congress authorized a task force, led by the Corps, to prepare a list of coastal wetland restoration projects in the state, and also provided funding to plan and carry out restoration projects in this and other coastal states under the Coastal Wetlands Planning, Protection and Restoration Act of 1990, also known as the Breaux Act.[38] The projects range from reintroduction of freshwater and diversion of sediment to construction of shoreline barriers and planting of vegetation. In total, the estimated total cost to complete all 147 approved projects is $1.78 billion.

In a 2007 report, GAO reported that it is impossible to determine the collective success of restoring coastal wetlands in Louisiana, because of an inadequate approach to monitoring. GAO had reviewed the Breaux Act program to identify the types of projects that have been designed and lessons that have been learned from 74 projects that have been completed so far.[39] Others, including the National Oceanic and Atmospheric Administration, disagreed with GAO's findings, observing that long-term data being provided through ongoing project monitoring are intended to yield insight into qualitative and quantitative project performance.

In the wake of hurricanes Katrina and Rita in 2005, multiple legislative proposals were introduced to fund additional restoration projects already planned by the U.S. Army Corps of Engineers and to explore other opportunities that would restore and stabilize wetlands in southern Louisiana. Before the hurricanes, Congress was considering legislation that would have provided about $2 billion to the restoration effort. Since the 2005 hurricanes, more expansive options costing up to $14 billion that were proposed in the 1998 report *Coast 2050* have also been considered.[40] The Gulf of Mexico Energy Security Act, legislation that authorizes additional revenues to states adjacent to offshore oil and gas production activities, was passed during the final days of the 109th Congress.[41] One of the purposes for which these revenues can be spent is wetland restoration, and the availability of these funds may affect the amount and scale of wetland restoration activity in the central Gulf Coast.

Concern for Louisiana's coastal wetlands was heightened by the oil spill following the April 2010 explosion of BP's drilling rig, the Deepwater Horizon, in the Gulf of Mexico. Although efforts focused on preventing oil from reaching coastal shorelines, some oil escaped capture and was pushed by wind and tides toward land. The degrees of impacts of oil on wetland vegetation are variable and complex and can be both acute and chronic, ranging from short-term disruption of plant functioning to mortality. The primary acute damage to the marshes is that plants, which hold the soil in place and stabilize shoreline, suffocate and die, especially if multiple coatings of oil occur. Once vegetation dies, the soil collapses. Then the soil becomes flooded, and plants cannot regrow. If plants cannot reestablish, soil erosion is accelerated, giving rise to even more flooding and further wetland loss. If oil penetrates into the sediments, roots are continuously exposed to oil, with chronic toxicity making production of new shoots problematic. Consequently, plant recovery is diminished, and eventually land loss occurs. In addition to direct impacts on plants, oil that reaches wetlands also affects animals that use wetlands during their life cycle, especially benthic organisms that reside in the sediments and are a foundation of the food chain.[42]

Public and private efforts were taken to protect the wetlands from oil that moved through Gulf waters towards coastal areas, but scientists remained concerned that high tides and wind could push oil into the marshes, and that the grasses and other vegetation that provide habitat for fish and wildlife would likely be destroyed. Wetland plants can be affected both by oil that floats over the surface of the marsh and by oil that has been incorporated into sediment. While oil was still flowing from the Deepwater Horizon site, cleanup of marshes was limited to triage of heavily oiled marshes and wetlands, because experts were concerned that greater harm than good could be done to the sensitive environmental ecosystems. The well was capped, and oil stopped flowing from the well site in mid-July 2010. Experts say that spill response efforts succeeded in keeping large amounts of oil from reaching coastal marshes. Nevertheless, oil remains in the Gulf environment, and potential for re-oiling of coastal areas, for example as a result of storms, will remain a concern for some time.

A recent federal report observes that Louisiana's *Coast 2050* is a comprehensive plan to protect and restore the state's coastal wetlands, but that other Gulf of Mexico states are only beginning similar planning processes for restoration of the damage caused by the Deepwater Horizon spill.[43]

WETLAND RESTORATION AND MITIGATION

Mitigation has become an important cornerstone of the Section 404 program in recent years. A 1990 MOA signed by the agencies with principal regulatory responsibilities (EPA and the Corps) outlines a sequence of three steps leading to mitigation: first, activities in wetlands should be avoided when possible; second, when they cannot be avoided, impacts should be minimized; and third, where minimum impacts are still unacceptable, mitigation is appropriate. Therefore, mitigation may be required as a condition of a Section 404 permit.

Federal wetland policies during the past 30 years have increasingly emphasized restoration of wetland areas. Much of this restoration occurs as part of efforts to mitigate the loss of wetlands at other sites. The mitigation concept has broad appeal, but implementation has left a conflicting record. Examination of this record, presented in a 2001 report from the National Research Council, found it to be wanting. The NRC report said that mitigation projects called for in permits affecting wetlands were not meeting the federal government's "no net loss" policy goal for wetlands function.[44] Likewise, a 2001 GAO report criticized the ability of the Corps to track the impact of projects under its current mitigation program that allows in-lieu-fee mitigation projects in exchange for issuing permits allowing wetlands development.[45] Both scientists and policymakers debate whether it is possible to restore or create wetlands with ecological and other functions equivalent to or better than those of natural wetlands that have been lost over time. Results so far seem to vary, depending on the type of wetland and the level of commitment to monitoring and maintenance.

Some wetland protection advocates are critical of mitigation, which they view as justifying destruction of wetlands. They believe that the Section 404 permit program should be an inducement to avoid damaging wetland areas. These critics also contend that adverse impacts on wetland values are often not fully mitigated and that mitigation measures, even if well-designed, are not adequately monitored or maintained. Supporters of current efforts counter that they generally work as envisioned, but little data exist to support this view. Questions about implementation of the 1990 MOA and controversies over the feasibility of compensating for wetland losses further complicate the wetland protection debate.

In response to criticism in the NRC and GAO reports on mitigation, in 2001, the Corps issued new guidance to strengthen the standards on compensating for wetlands lost to development. But the guidance was criticized by environmental groups and some Members of Congress for weakening rather than strengthening mitigation requirements and for the Corps' failure to consult with other federal agencies. In 2002, the Corps and EPA released an action plan including 17 items that both agencies believed would improve the effectiveness of wetlands restoration efforts.[46]

In 2008, the Corps and EPA promulgated a mitigation rule to replace the 1990 MOA with clearer requirements on what will be considered a successful project to compensate for wetlands lost to activities like construction, mining, and agriculture.[47] The rule sets performance standards and criteria for three types of wetlands mitigation: mitigation banks, in-lieu programs, and permitteeresponsible compensatory mitigation. It sets standards to mitigate the loss of wetlands and associated aquatic resources and is intended to improve the planning, implementation, and management of compensatory mitigation projects designed to restore aquatic resources that are affected by activities that disturb a half-acre or more of wetlands. It also is designed to help ensure no net loss of wetlands by addressing key recommendations raised in the 2001 NRC report. Under the rule, all compensation projects must have mitigation plans that include 12 fundamental components, such as objectives, site selection criteria, a mitigation work plan, and a maintenance plan.[48]

The concept of "mitigation banks," in which wetlands are created, restored, or enhanced in advance to serve as "credits" that may be used or acquired by permit applicants when they are required to mitigate impacts of their activities, is widely endorsed and is the preferred option under the 2008 mitigation rule. Numerous public and private banks have been established, but many believe that it is too early to assess their success. In a study of mitigation, the Environmental Law Institute determined that as of 2005, there were 330 active banks, 75 sold-out banks, and 169 banks seeking approval to operate.[49] More recent data paint a larger picture. According to the Army Corps, as of November 2014, there are 1,408 approved and active mitigation banks, 276 banks seeking approval, and 221 sold-out banks.[50]

Congress has repeatedly endorsed mitigation through legislation. Provisions in several laws, such as the farm bill and the 1998 Transportation Equity Act (TEA-21), endorse the mitigation banking concept. In 2003, Congress enacted wetlands mitigation provisions as part of the FY2004 Department of Defense (DOD) authorization act (P.L. 108-136). Section 314 of that act directed DOD to make payments to wetland mitigation banking programs in instances where military construction projects would result or could result in destruction of or impacts to wetlands.

End Notes

[1] Two places to view material on some of the changes in scientific knowledge and understanding are through the products of the Association of State Wetlands Managers, http://www.aswm.org, and the Society of Wetland Scientists, http://www.sws.org

[2] Thomas E. Dahl, Status and Trends of Wetlands in the Conterminus United States, 2004-2009, U.S. Department of the Interior, Fish and Wildlife Service, 2011.

[3] For information, see http://www.fws.gov/wetlands/Status-And-Trends/index.html.

[4] Thomas E. Dahl and Susan-Marie Stedman, Status and Trends of Wetlands in the Coastal Watersheds of the Conterminous United States 2004-2009, U.S. Department of the Interior, Fish and Wildlife Service, and National Ocean and Atmospheric Administration, National Marine Fisheries Service, October 2013.

[5] See http://water.epa.gov/type/wetlands/assessment/survey/index.cfm.

[6] Office of the President, Council on Environmental Quality, Conserving America's Wetlands 2008: Four Years of Progress Implementing the President's Goal, April 2008.

[7] T. E. Dahl, Status and Trends of Wetlands in the Conterminous United States, 2004-2009, U.S. Department of the Interior, Fish and Wildlife Service, 2011, p. 86.

[8] Matthew L. Kirwan, Glenn R. Guntenspergen, and Andrea D'Alpaos et al., "Limits on the adaptability of coastal marshes to rising sea level," Geophysical Research Letters, vol. 37, no. L23401 (December 2010).

[9] While tidal wetlands do effectively sequester carbon, some wetlands (especially those with low salinity levels) also are a source of GHGs by emitting methane, which is approximately 21 times more powerful as a GHG than CO2.

[10] Stephen Crooks, Dorothee Herr, and Jerker Tamelander et al., "Mitigating Climate Change Through Restoration and Management of Coastal Wetlands and Near-shore Marine Ecosystems, Challenges and Opportunities," The World Bank Environment Department,

March 2011, http://siteresources.worldbank.org/ENVIRONMENT/Resources/ MtgtnCCthru MgtofCoastalWetlands.pdf.

[11] Stephen Emmett-Mattox, Stephen Crooks, and Jette Findsen, "Wetland Grasses and Gases: Are Tidal Wetlands Ready for the Carbon Markets?," National Wetlands Newsletter, November-December 2010, pp. 6-10.

[12] For additional information, see CRS Report RL31411, Controversies over Redefining "Fill Material" Under the Clean Water Act, by Claudia Copeland.

[13] The updated list of wetland plants is available at http://wetland_plants.usace.army.mil.

[14] For information, see CRS Report 97-223, The Army Corps of Engineers' Nationwide Permits Program: Issues and Regulatory Developments, by Claudia Copeland

[15] For additional information, see CRS Report RL30849, The Supreme Court Addresses Corps of Engineers Jurisdiction Over "Isolated Waters": The SWANCC Decision, by Robert Meltz.

[16] See http://www.epa.gov/owow/wetlands/pdf/Joint_Memo.pdf.

[17] U.S. Government Accountability Office, Corps of Engineers Needs to Evaluate Its District Office Practices in Determining Jurisdiction, GAO-04-297, February 2004.

[18] U.S. Congress, House of Representatives, Committee on Transportation and Infrastructure, Subcommittee on Water Resources and Environment, Inconsistent Regulation of Wetlands and Other Waters, Hearing 108-58, 108th Cong., 2nd sess., March 30, 2004.

[19] For additional information, see CRS Report RL33263, The Wetlands Coverage of the Clean Water Act (CWA): Rapanos and Beyond, by Robert Meltz and Claudia Copeland.

[20] U.S. Environmental Protection Agency and Army Corps of Engineers, "Draft Guidance on Identifying Waters Protected by the Clean Water Act," April 27, 2011, p. 2, on file with author.

[21] For further discussion, see CRS Report R43455, EPA and the Army Corps' Proposed Rule to Define "Waters of the United States", by Claudia Copeland.

[22] For information on the 111th Congress legislation, see CRS Report R41225, Legislative Approaches to Defining "Waters of the United States", by Claudia Copeland.

[23] Nancy Sutley, Chair, Council on Environmental Quality, et al., letter to Senator Barbara Boxer, Chair, Senate Environment and Public Works Committee (and other congressional leaders), May 20, 2009, http://epw.senate.gov/public/index.cfm?FuseAction=Majority.Press Releases &ContentRecord_id=64739ae3-802a-23ad-4c30-36fc58cc1014& Region_id=&Issue_id=.

[24] For information, see CRS Report R40763, Agricultural Conservation: A Guide to Programs, by Megan Stubbs.

[25] For details, see CRS Report R43504, Conservation Provisions in the 2014 Farm Bill (P.L. 113-79), by Megan Stubbs.

[26] For additional information, see CRS Report R40763, Agricultural Conservation: A Guide to Programs, by Megan Stubbs.

[27] See http://www.usace.army.mil/Portals/2/docs/civilworks/mous/foodsecurity_cleanwateract.pdf.

[28] See http://www.fws.gov/partners/.

[29] For information, see http://www.fws.gov/coastal/CoastalGrants/index.html.

[30] See http://www.fws.gov/birdhabitat/Grants/NAWCA/index.shtm.

[31] U.S. Congress, House of Representatives, Committee on Transportation and Infrastructure, Subcommittee on Water Resources and Environment, The Wetland Permitting Process: Is It Working Fairly? Hearing 107-50, 107th Cong., 1st sess., October 3, 2001.

[32] Environmental Law Institute, State Wetland Protection: Status, Trends & Model Approaches, March 2008, p. 6. Hereinafter, ELI State Wetland Protection.

[33] See Association of State Wetland Managers, "State Wetland Program Summaries," at http://www.aswm.org/statesummaries.

[34] Twenty-two states rely on Section 401 as the sole form of state-level regulation, and 15 additional states rely on Section 401 as the primary form of state-level regulation but also have adopted laws that provide additional protection to certain wetland categories, such as coastal wetlands. ELI State Wetland Protection, p. 13.

[35] See Environmental Council of the States, "The States' Definitions of 'Waters of the State,'" February 2009, at http://www.ecos.org/section/publications.

[36] ELI State Wetland Protection, p. 17.

[37] B. R. Couvillion, J. A. Barras, and G. D. Steyer, et al., Land Area Change in Coastal Louisiana from 1932 to 2010, U.S. Geological Survey, Pamphlet to accompany U.S. Geological Survey Scientific Investigations Map 3164, June 2011, http://pubs.usgs. gov/sim/3164/ downloads/SIM3164_Pamphlet.pdf.

[38] For information on this program, see CRS Report RS22467, Coastal Wetlands Planning, Protection, and Restoration Act (CWPPRA): Effects of Hurricanes Katrina and Rita on Implementation, by Jeffrey A. Zinn.

[39] U.S. Government Accountability Office, Coastal Wetlands: Lessons Learned from Past Efforts in Louisiana Could Help Future Restoration and Protection, GAO-08-130, December 2007.

[40] See http://www.coast2050.gov. For a more detailed discussion of the effects of the hurricanes on planning for wetland restoration, see CRS Report RS22276, Coastal Louisiana Ecosystem Restoration After Hurricanes Katrina and Rita, by Jeffrey A. Zinn.

[41] S. 3711 was attached to a broad tax relief measure that was enacted in December 2006 (H.R. 6111, P.L. 109-432). For additional information, see CRS Report RL33493, Outer Continental Shelf: Debate Over Oil and Gas Leasing and Revenue Sharing, by Marc Humphries.

[42] Dennis F. Whigham, Stephen W. Broome, and Curtis J. Richardson et al., Statement of the Environmental Concerns Committee, Society of Wetland Scientists, "The Deepwater Horizon Disaster and Wetlands," http://www.docsrush.net/3039416/the-deepwater-horizon-disaster-and-wetlands-statement-from.html.

[43] Thomas E. Dahl and Susan-Marie Stedman, Status and Trends of Wetlands in the Coastal Watersheds of the Conterminous United States 2004-2009, U.S. Department of the Interior, Fish and Wildlife Service, and National Ocean and Atmospheric Administration, National Marine Fisheries Service, October 2013, p. 37.

[44] National Academy of Sciences, National Research Council, Compensating for Wetland Losses under the Clean Water Act (Washington, DC, 2001).

[45] U.S. Government Accountability Office, Wetlands Protection: Assessments Needed to Determine the Effectiveness of In-Lieu-Fee Mitigation, GAO-01-325, May 2001.

46 U.S. Environmental Protection Agency and U.S. Army Corps of Engineers, "National Wetlands Mitigation Action Plan, December 24, 2002." See http://water.epa.gov/ lawsregs/guidance/wetlands/upload/ 2003_07_10_wetlands_map1226withsign.pdf.

[47] U.S. Army Corps of Engineers and Environmental Protection Agency, "Compensatory Mitigation for Losses of Aquatic Resources, Final Rule," 73 Federal Register 19594, April 10, 2008.

[48] Information on compensatory mitigation can be found at http://water.epa.gov/ lawsregs/ guidance/wetlands/ wetlandsmitigation_index.cfm.

[49] For more information on mitigation generally, and mitigation banks specifically, see Environmental Law Institute, 2005 Status Report on Compensatory Mitigation in the United States, April 2006.

[50] See http://geo.usace.army.mil/ribits/index.html.

In: Wetlands
Editor: Nelson E. Santiago
ISBN: 978-1-63482-728-7

Chapter 2

TARGETING INVESTMENTS TO COST EFFECTIVELY RESTORE AND PROTECT WETLAND ECOSYSTEMS: SOME ECONOMIC INSIGHTS*

LeRoy Hansen, Daniel Hellerstein, Marc Ribaudo, James Williamson, David Nulph, Charles Loesch and William Crumpton

ABSTRACT

USDA has spent more than $4.2 billion on wetland restoration and protection over the last two decades. One challenge in allocating these funds is the lack of information on variations in wetland benefits and costs across the Nation. This report discusses the biophysical impacts of new wetlands for eight benefit categories: duck hunting, carbon sequestration, flood protection, nitrogen removal, species protection, open space, sediment removal, and groundwater recharge, as well as the value of these impacts for some categories. In addition, it presents county-level estimates of the costs of restoring and preserving wetlands

* This is an edited, reformatted and augmented version of Economic Research Report Number 183, issued by the U.S. Department of Agriculture, Economic Research Service, February 2015.

for some parts of the United States. Although the estimates range in precision and are not comprehensive, they call attention to some areas where the benefits of new wetlands are likely to exceed costs or perhaps may be insignificant. For example, the benefits of restoring and preserving wetlands near the Missouri River in central North and South Dakota are likely to exceed costs. Findings underscore the need for additional information that may increase the number, accuracy, and spatial resolution of wetland benefit estimates.

Keywords: Wetland conservation, ecosystem benefits, environmental economics, environmental targeting, nonmarket benefits, values of wetland ecosystems

What Is the Issue?

Over the last several decades, concerns over the loss of wetland ecosystems have led to legislation restricting wetland conversion and promotion and funding of wetland restoration and conservation. Consequently, USDA has spent more than $4.2 billion on wetland restoration and protection during the period. The environmental benefits of wetland conservation—cleaner water, increases in wildlife populations, and carbon sequestration—depend on climate, human influences, and landscape characteristics. This study assesses how future wetland conservation funding might be targeted within States and regions and across the United States to maximize benefits relative to costs of restoring and preserving wetlands. Information on the sizes and spatial distributions of the benefits and costs of wetland protection can help public and private decisionmakers understand the economic implications of efforts to reduce wetland losses, improve existing wetlands, and restore prior-existing wetlands through regulation and economic incentives.

What Did the Study Find?

A number of different types of wetland benefits were considered as part of this analysis. The ranges in wetland benefits and costs across the country demonstrate potential gains from targeting public and private wetland restoration efforts by location and by wetland characteristics. Some benefits can be estimated more precisely than others:

- The annual benefits to duck hunting from restoring wetlands in the Prairie Pothole Region (primarily, lands in the eastern Dakotas and western edges of Minnesota and Iowa) range from near $0 to $143 per wetland acre.
- The annualized value of a new wetland's effect on atmospheric concentrations of greenhouse gases ranges from $0 to $129 per wetland acre in the areas studied. In other areas, wetlands can increase, decrease, or have no net effect on atmospheric levels of greenhouse gases.
- Wetlands on agricultural land are not likely to provide notable flood-protection benefits unless they are located in close proximity to urban areas.
- Wetlands in the Upper Mississippi and Ohio River watersheds can annually remove from 11 to 1,800 pounds of nitrogen per wetland acre in waters leaving farm fields from watershed streams and rivers.
- About 1,756 counties in the contiguous States have wetland-associated imperiled species, with some counties having as many as 23 imperiled species. About 1,355 counties have no wetland-associated imperiled species, but this does not suggest that the value of wetland-species protection benefits in these counties is low. The analysis found no information that illustrates how a new wetland might affect species' survival probabilities. Additionally, it found no information on the public's willingness to pay (or how it might prioritize support) for changes in imperiled species' survival probabilities.
- Many wetlands recharge groundwater supplies, but the majority do not. Some wetlands exist because they are fed by groundwater. Others exist because they have impermeable subsoil.
- Sediment removed by wetlands, while protecting other bodies of water, can reduce other wetland benefits as the sediment accumulates.
- The benefits of wetlands as open-space protection in rural areas are likely to be very small.

Based on contract data from USDA's Wetlands Reserve Program, the costs of restoring and preserving a wetland range from $170 per acre in the western Dakotas, Montana, Arkansas, and Louisiana to $6,100 per acre in the major corn-producing areas and along the Northern Pacific Coast. These estimates reflect the change in the value of the land with and without a permanent wetland easement plus wetland restoration costs.

An analysis of the combined benefits and costs of new wetlands revealed the following:

- Restoring wetlands in the western Prairie Pothole Region is likely to generate duck-hunting benefit-cost ratios close to or greater than one; that is, restoration funds spent in this area would be a good investment. However, most of the people who would benefit from the added duck populations in the region are hunters who live elsewhere.
- The value of new wetlands' greenhouse gas impacts exceeds wetland costs in the western half of the Prairie Pothole Region and in most of the Mississippi Alluvial Valley (from southern Illinois, covering western portions of Tennessee and Mississippi and eastern portions of Missouri and Louisiana).
- The amount of nitrogen that can be removed for each dollar spent on wetland restoration and protection in the Upper Mississippi and Ohio River watersheds ranges from 0.2 to 34.0 pounds. The costs of nitrogen removal through new wetlands range from $0.03 to over $5.00 per pound of reduced N loadings in nearby waterways.

How Was the Study Conducted?

To quantify wetland ecosystem effects on amenities, and the value the public places on these amenities, ERS researchers reviewed and analyzed economic, ecological, and engineering literature and data. In some cases, researchers used existing data and models; new models were also developed. Benefits were estimated at the county and subcounty levels across limited portions of the Nation, depending on the spatial resolution (e.g., geographic reliability) of supporting data and economic and ecological models.

Researchers used Wetlands Reserve Program contract data to empirically estimate wetland cost models for major wetland regions. These models were then used to generate county-level estimates of the costs of new wetlands. Where the estimates allowed, researchers generated county- and subcounty-level benefit-cost ratios by dividing the benefit estimates by the cost estimates. These ratios illustrate areas where wetland restoration funding might be allocated to best serve public interests. When it was not possible to estimate economic benefits, researchers generated ratios of the wetlands' cost effectiveness—the quantity of services produced relative to costs. Lastly, where data were limited, representative measures were generated that illustrate the scope and the magnitude of possible benefits.

INTRODUCTION

Since European settlement, the 48 contiguous States are estimated to have lost about half of the estimated 220 million original wetland acres. Some areas lost greater shares than others—regions within Iowa, Illinois, and Indiana lost as much as 90 percent. Most losses stemmed from drainage for agriculture, as farmers found that the land provided returns that covered the cost of wetland conversion. In fact, prior to the 1970s, the Federal Government provided subsidies for some wetland conversion.

In more recent years, the breadth of ecological services provided by wetlands, ranging from serving as critical habitat of many plant, fish, and wildlife species to carbon sequestration, has been widely recognized and appreciated. Nonetheless, economic incentives for private landowners to provide most wetland services are generally weak—the benefits provided by wetlands and their ecosystems often have a nonmarket "public goods" nature. Furthermore, the individuals that benefit may be far from the wetland, making the collection of monetary returns all the more difficult.

Public goods are characterized as being nonrival in consumption (an additional individual can enjoy the good without affecting others) and by the difficulty of excluding nonpayers (Kaul et al., 1999). For example, the view of a wetland from public land, such as a highway, is a pure public good. Similarly, wetlands can boost duck populations. But it is difficult to exclude those who are not willing to pay for the subsequent increase in hunting quality and views of migratory ducks. Generally speaking, even though some amenities provided by wetlands are not pure public goods, the public-good characteristics (as well as wetlands' effects on other amenities) are likely to make market provision inefficient (Ribaudo et al., 2008).

Since the 1970s, having recognized the public goods provided by wetlands, Federal, State, and local governments, as well as nongovernmental organizations (NGOs), have initiated efforts to reduce wetland losses, improve existing wetlands, and restore prior-existing wetlands through regulation and economic incentives. USDA's most significant wetland restoration effort was the Wetlands Reserve Program (WRP), a voluntary program established by the Food, Agriculture, Conservation, and Trade Act of 1990 (the 1990 Farm Act).[1]

The objective of the WRP was to protect, restore, and enhance wetland ecosystems to protect and improve:

- Habitat for migratory waterfowl and other wetland-associated wildlife, including imperiled species
- Water quality and groundwater recharge
- Floodplains and water flows to reduce flooding
- Aesthetics and open space, and native flora and fauna
- Education and scientific scholarship

By the end of 2012, approximately 2.6 million acres were enrolled in the WRP, of which 2.1 million acres are restored wetlands, about 0.14 million acres are prior-existing wetlands, and the remainder are surrounding uplands.[2] The WRP's financial commitments varied over time but averaged about $200 million per year from 2009 to 2012.

The Conservation Reserve Program (CRP) is USDA's other major wetland conservation effort, though wetland conservation is not the primary goal of the program. By the end of 2012, the CRP had 2.29 million wetland-related acres under contract, including 0.50 million acres of improved, prior existing wetlands; 0.54 million acres of newly restored wetlands; and 1.251 million upland acres (e.g., dry lands surrounding wetlands). In 2012, annual expenditures on CRP wetland-related contracts totaled about $175 million.[3] Together, the WRP and CRP protect about 2.6 million restored and 6.4 million prior-existing wetland acres.[4]

Four features in the design of the WRP and CRP are noteworthy, particularly in regards to wetland conservation programs and NGO initiatives. First, for both the WRP and CRP, a participating land- owner maintains ownership of the land, is allowed limited uses of the land, and can sell the land. However, the contractual agreement, and its associated constraints, stays with the land in the event of a transfer of ownership.

Second, the lengths of the contracts vary by program. About 87 percent of the WRP contract acres are permanent easements, 6 percent are 30-year easements, and the remainder are 10-year agreements. In contrast, CRP wetland-related contracts range from 10 to 15 years (about 25 percent are for 10 years, 75 percent are for 14 or 15 years). The differences in contract lengths may affect the types of benefits generated. For example, some wetland ecosystems take years to recover. In such cases, the annual services of newly restored or improved wetlands might increase over time as they move toward their full potential.

Third, the program payment schedules vary by program. In the WRP, a landowner receives a one- time upfront payment from the Federal Government. In contrast, CRP contracts are agreements under which the landowner receives annual payments. This design feature engenders comparisons of benefits and costs provided by the two programs over time. The WRP is an investment that provides wetland benefits, in most cases indefinitely, whereas the annual CRP payments are more analogous to annual rental payments, and the stream of environmental benefits flowing from the retired land may decline after the contract expires. From 2009 to 2013, approximately 50 percent of the CRP acres in expiring contracts were re-enrolled.[5] For the other 50 percent, historical use suggests that a portion of the land might return to agricultural production—these lands were typically in crop production in the 1980s and, while profitable then, recent increases in commodity prices have boosted returns. But under the following circumstances, landowners may choose not to drain wetlands: the land is more profitable if left as a wetland (i.e., it may be profitable to allow forested wetlands to mature); the landowner enjoys having a wetland; or the costs to covert the land are prohibitively high. And, if wetlands are left undrained, wetland ecosystems will continue to produce public benefits without public funding.

Lastly, competitive enrollments can reduce program costs and increase benefits. Generally, lands that can provide wetland benefits are accepted in the CRP as long as they meet certain baseline requirements, such as having been used in crop or livestock production.[6] Enrollment in the WRP is competitive. To participate, landowners submit offers to USDA's Natural Resources Conservation Service (NRCS) State field offices. The State field offices assign scores to each offer based on ranking criteria. Most State offices develop their own ranking criteria, though some States have worked together. The ranking criteria award points to contract offers based on a range of expected environmental effects and, in some States, costs. Within each State, offers receiving the highest scores are generally accepted, given budget constraints and acreage enrollment targets.

Federal, State, and local governments and NGOs that wish to develop and improve ranking criteria that represent public interests in wetlands' environmental impacts face at least two major challenges. First, they must determine how a new wetland and its ecosystem might affect environmental amenities. Second, they must assign weights to the various wetland amenities that reflect the relative strengths of public preferences. Because the information needed to make these determinations is limited, ranking criteria are based on the judgments of program managers, individuals, community groups, NGOs, and commercial enterprises.

This study assesses how future wetland conservation funding might be targeted within States, regions, and across the United States to maximize benefits relative to costs. It uses data, model results, and findings from previous research to generate spatial estimates of the costs and benefits of new wetlands. That is, to the extent possible, the analysis taps biophysical and economic sciences to quantify wetlands' and their ecosystems' effects on amenities, the public's preferences for amenities, and wetland costs. Findings offer geographically disaggregate illustrations of costs and some environmental benefits and impacts in some regions of the United States. Findings also illustrate the many areas where more information is needed.

THE COSTS OF WETLAND ESTABLISHMENT

The primary data used in the cost analyses come from WRP contracts. The WRP file includes the amounts of the payments accepted by landowners for the easement and wetland restoration—the dependent variable in the cost functions. The WRP contract data are particularly appealing because the program, by congressional design, caps easement payments at the difference between the values of land before and after it is enrolled. When administered correctly, easement payments are not likely to overcompensate landowners and are a good measure of opportunity costs. WRP data do not include the cost of technical support provided by USDA. Thus, the cost estimates may understate the total cost of restoring and protecting a wetland. Any bias in the estimates, however, is expected to be relatively small.

Clearly, larger wetland easements will require greater easement (incentive) payments and have greater restoration costs, all else being equal. But a number of other factors affect the size of the payments landowners are willing to accept and the wetland restoration costs. The WRP data provide other variables, including a county identifier, which enable one to link each observation to other county-level data.

Drivers of Wetland Easement and Restoration Costs

Restoring and preserving a wetland through a publicly funded voluntary program involves two types of costs: the net opportunity cost (current minus the alternative land use values) is the cost of the permanent wetland easement;[7] and the cost of restoring wetland hydrology can include simple measures (blocking or slowing water movement) or significant construction (dredging wetland fill), as well as the establishment of appropriate flora and wildlife habitat in the wetland.

Given the program's features and based on a review of the literature, the biggest drivers of wetland *easement* costs are the following:

1. *Current land value*—Landowners will want to be compensated for what they give up, which will usually be less than the market value of their land, given that the remaining property rights have value. The current market value of farmland is at least as great as its agricultural value but can be higher when there is a potential for urban or other higher valued uses.
2. *The value of the land with an easement*—The amount that a landowner gives up when selling the easement (e.g., the opportunity cost of the easement) also depends on the value of the land with the easement—the compensation payment a landowner demands will be inversely related to the value of the land with the easement. The market value of land with an easement will be at least equal to its remaining agricultural-use value (i.e., occasional grazing and haying) plus the value of the land for onsite recreation (i.e., bird watching and hunting), but it could be greater. It is expected that the remaining agricultural value will be correlated with the size of the easement and, possibly, its pre-wetland agricultural value. The onsite recreation value is expected to be directly affected by the number of people likely to use the wetland, which is likely to be closely related to the size of the surrounding population.[8]
3. *The value of land near the wetland*—Landowners who own other land near the wetland are likely to consider the effects of the wetland on the values of those lands—should the value of surrounding lands increase (decrease), the landowner is likely to accept a lower (higher) easement payment. Findings suggest that the value of land surrounding a wetland easement can increase when buyers are willing to pay a premium for land with a wetland view (Ma and Swinton, 2011; Mahan et al., 2000; Doss and Taff, 1996) —the greater the urban pressure, the greater the premium. Note that Reynolds and Regalado (2002) found that the type of wetland affects the premium. But findings also suggest that a restored wetland nearby can reduce

the urban development value of land. In an analysis of the rapidly growing Raleigh-Durham- Chapel Hill region of North Carolina, Kazak and BenDor (2013) found that a new wetland lowered the urban-development value of land within 0.5 miles of the wetland. While urban development pressure can have some effect, the direction of the effect, if any, is uncertain.

4. *Nuisance cost*—Nuisance cost is the increase in fuel, time, and fencing needed to maneuver equipment around and keep livestock out of new wetlands—or, alternatively, the opportunity cost of altering farm production or cattle grazing on the surrounding lands. The nuisance cost is borne by the landowner. Easement payments will need to offset nuisance costs to encourage landowner participation (Pattison et al., 2011). It is expected that larger contracts will impose greater nuisance costs, driving up the cost of an easement.
5. *Good steward*—A landowner wishing to be a "good steward" may accept a lower easement payment. While good stewardship can have an effect on the easement payment a landowner would find acceptable, this analysis does not include an explicit measure of such.

The primary drivers of (or factors affecting) wetland *restoration* costs are the following:

1. *Size*—Larger easements require more restoration resources.
2. *Hydrology*—The cost of restoring a wetland's hydrology depends on the level of construction needed. For example, filling channels, eliminating tile drainage, and restoring flow impediments can be much less costly than excavating filled wetlands. The actions taken to restore wetlands' hydrology would not be expected to differ substantially across wetlands of the same type.
3. *Flora and fauna*—The cost of restoring flora and fauna that support wetland ecosystems depends on the type of activities involved, such as planting seeds or planting live vegetation, transporting live fauna or allowing natural recovery, planting by machine or by hand, and preventing or controlling invasive species. For example, restoring grassed wetland may only require drill-seeding the land prior to restoring the hydrology, while restoring a forested wetland may require planting young plants by hand and, hence, may be very labor

intensive. The actions taken to restore wetlands' fauna would not be expected to differ substantially across wetlands of the same type.

Based on these drivers, a database was developed to serve as the explanatory variables (including vari able proxies) in a model of wetland costs. We first considered current farmland values. Because spatial data on farmland values nationwide are not available, we used county-level data on farmland rental rates—which are, effectively, an annualized measure of the agricultural value of land. We therefore generated a proxy variable for the agricultural value of land in an easement by multiplying the rental rate values by the number of acres in the easement. Note that, as discussed earlier, the agricultural value of land is expected to have positive effects on lands with and without wetland easements.

Another variable included in the database serves as an indicator of urban pressure. Urban pressure can drive up farmland values—the greater the potential for nonfarm development, the greater the land value. But urban pressure can also drive up the values of lands with easements and lands surrounding easements. Hence, the net effect of urban pressure is uncertain.

A third variable included is the size of the wetland. For a given type of wetland, the cost of restoring the wetlands' hydrology and ecosystems and nuisance costs are expected to increase with the size of the wetland. The agricultural value of land with a wetland is also expected to increase with easement size, which may partially offset the other two effects. However, there may be economies of scale or fixed costs associated with restoration and the other factors so that costs may not be proportional to the size of the wetland. To account for nonlinear relationships, a higher order term of this variable was included in the regression models.

To control for differences in restoration costs across wetland types, separate models were estimated for each of 10 major U.S. wetland regions, as defined by NRCS, where adequate data are available (figure 1). By design, these regions reflect wetland similarities. It seems reasonable to assume that, within a wetland region, determinants of the cost of restoring a wetland—other than the size of the wetland—are likely to be similar. The demand for the view of a wetland will depend on the appearance of the wetland and the type of ecosystem it supports, which is expected to show little spatial variation for a given type of wetland. Regional analysis helped control for several drivers of the cost of wetland easements. (For details on the model variables, see appendix 1.[9])

USDA, Natural Resources Conservation Service.

Figure 1. Major wetland regions of the contiguous States.

Predicting the Public Cost of Establishing New Wetlands

Across the 10 models, most coefficients are statistically significant at the 99-percent level, and the estimated models explain 53 to 88 percent of the variation in wetland costs (as measured by each model's adjusted R-square). The five best-fitting wetlands cost models capture more than 75 percent of the variation in total wetland cost, and the three weakest capture 53 to 60 percent. (For estimation results, see appendix 1.)

Using these results to generate county-level estimates of the total cost of a new wetland requires selecting appropriately representative sizes for each wetland region. For this analysis, we chose the median size of previously established WRP contracts (within a region), as this facilitates generating an estimate of median cost. Given the nonlinearity of the cost models, the average-sized contract will not generate an estimate of the average contract cost. We also used county-specific measures of agricultural land rental rates. Thus, the county estimates are driven by a regionally representative median contract size, the region-specific estimated coefficients, and county-specific agricultural land rental value.

To allow cross-region comparisons, we converted total wetland cost estimates to per-acre costs. The per-acre cost estimates range from $170 to $6,100, with some of the lowest costs in western North Dakota and eastern Montana and the highest in major corn-producing areas and western Washington and Oregon (figure 2).[10] Wetland costs are most variable spatially within the Prairie Pothole Region (where values increase from northwestern to southeastern extremes), Northwest (where values are high in much of the region), and Appalachian Plains (where values vary sharply throughout). Wetland costs could not be estimated for areas that lie outside the NRCS wetland regions examined in this analysis.

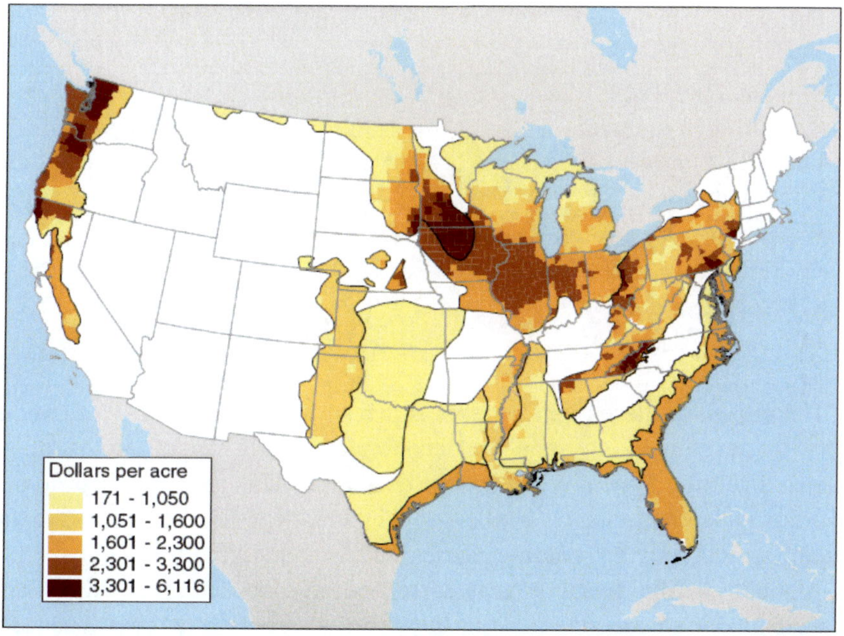

White denotes areas where cost estimates were not calculated. The major boundary lands on the map define the wetland regions shown in figure 1.

Source: USDA, Economic Research Service using data from Wetlands Reserve Program contract files.

Figure 2. County-level estimates of the cost of restoring and preserving a new wetland.

SELECTED ESTIMATES OF THE BENEFITS OF WETLAND CONSERVATION

Wetlands have been credited with increasing and enhancing water quality, protecting fish and wildlife populations, providing flood protection, sequestering carbon, and providing open space and aesthetic beauty.[11] This section briefly describes the biophysical and economic models used to estimate wetland services, amenities, and the value of amenities.[12] We then present a summary of findings on the values of a suite of eight types of wetland benefits within a number of wetland regions. The estimates of wetland benefits are reported in annual values. To provide a more intuitive comparison of costs and benefits, we convert wetland costs to annualized values using a 4-percent discount rate.[13]

The eight wetland benefits examined include the following:

- Duck hunting
- Nitrogen removal
- Species protection
- Flood protection
- Greenhouse gas reductions
- Groundwater recharge
- Sediment removal
- Open space

The choice of these eight benefits is driven by the availability of data, the ability to develop biophysical and economic models, and findings of past research that have some level of spatial resolution. This group reflects some of the often-stated benefits of wetlands. Wetland benefits not addressed here include impacts on other wildlife, increased biodiversity, reduced phosphorous and pesticide loadings, and improved recreational and commercial fisheries. We were unable to generate most of the benefits across all regions. The results presented here do not reveal the optimal allocation of U.S. wetland restoration funds because the number of benefits estimated is limited and confined to only parts of the United States. Even so, the estimates (1) provide a perspective on the size and spatial variation in some benefits in some areas; (2) can aid judgments on how to weave reported and unreported economic and biophysical information together to support information-based decisions on targeting wetland conservation funding; and (3) illustrate the need for additional data and analyses to further support information-based decisionmaking.

Estimating the Value of Wetlands and Their Ecosystem Services

To estimate the value of a new wetland's benefits, one must apply economic and biophysical sciences. Biophysical sciences enable one to quantify the services that a new wetland and its ecosystems are likely to produce. The same sciences can link changes in services to changes in environmental amenities—that is, the things people care about. Economic analyses can enable one to estimate the public's willingness to pay for changes in amenities.[14]

Each of these economic and biophysical tasks is critical; the strength of a wetland benefit estimate is only as strong as its weakest link. As one's understanding of each link improves, so, too, will the analyses of wetlands' spatial benefits.[15] Furthermore, a better understanding of these links will help one better understand the tradeoffs embedded in wetland ranking criteria.

Table 1 lists some of the services wetlands and their ecosystems produce, the amenities that services produce, and the ways in which the public values amenities. Appendix 2 provides details on the biophysical and economic framework critical in evaluating the benefits of a new wetland.

This framework for estimating the value of benefits requires several kinds of information (both data and models). The specific units that define a wetland service and the amenities provided by each service are "grey" in that the way in which they are specified by an analyst will depend on the available data, models, and other information. What is important is the conceptual framework that the functions illustrate. When this information is not fully available, a number of shortcuts can be taken, including reduced-form models, proxy variables, and meta analysis. But, as discussed in appendix 2, several features leave meta analyses unsuited to generate spatially disaggregated estimates of benefits.

Duck Hunting Benefits: What the Study Found

Wetlands directly affect duck populations. Between 55 and 80 percent of all North American ducks nest in the Prairie Pothole Region (PPR) (see figure 1). Ducks of the PPR region migrate using all of the continent's major flyways, including the coastal flyways. Increases in duck populations generate a variety of amenities that people throughout the country enjoy:

- Improved hunting quality
- Better bird watching throughout the year
- An increase in the size and number of fall migrating flocks

This study focuses on wetlands' impacts on hunting quality and people's willingness to pay for the changes.

Table 1. Functional relationships between new wetlands' direct effect on the environment and the value the public places on the subsequent impacts on amenities

Biophysical and environmental processes			Economic valuations
Resource	***The services produced by wetland x***	***Amenities that stem from services***	***The public's demands or willingness to pay for amenities***
ECOSYSTEM SERVICES	$fi(x) \rightarrow$	$gj(fi(x)) \rightarrow$	$Dk(gj(fi(x)))$
	Habitat for terrestrial and aquatic species	Improved species' survival rates	What are people willing to pay to better protect species?
		Improved duck hunting quality	What are people willing to pay for better hunting?
		Increased health of game fish species	What are people willing to pay for better recreational fishing?
		Increased health of commercial fish species	What is the change in the commercial fishing industry's profits?
	Nutrient removal	Less algae in freshwaters	What are people willing to pay to have cleaner water?
		Commercial fish popula- tions less affected by hypoxic zones	What is the change in the commercial fishing industry's profits?
		Lower nitrogen level in waters used by municipalities	What are the municipal water-treatment costs savings associated with reduced nitrogen in intake waters?

Biophysical and environmental processes			**Economic valuations**
Resource	***The services produced by wetland x***	***Amenities that stem from services***	***The public's demands or willingness to pay for amenities***
	Reductions in greenhouse gas	Economic, ecological, and health impacts of climatic change	What are people willing to pay for a reduction in the economic, ecological, and health impacts of climatic change?
SERVICES	Groundwater recharge	Drinking water supply	What is a municipality willing to pay for an increase in water supply?
		Groundwater use in agriculture	How much will increases in groundwater affect agricultural profits?
		Groundwater's effect on stream amenities	What are people willing to pay to see an increase in the amenities?
	Floodwater retention	Reduced flood heights (wetland bounce) and water speeds	What is the expected value of the reduced expected property damages?
			What are people willing to pay to reduce risks to health and safety?
	Open space	Improved views	What are people willing to pay to view a wetland, given its location?

Source: USDA, Economic Research Service.

For this analysis, we developed a biophysical model and applied findings of earlier research to generate spatial estimates of the effects of new wetlands on duck populations (an ecosystem service). Then, using findings of past research and reasonable assumptions, we linked changes in duck populations to changes in hunting quality (an amenity). Finally, we used benefits transfer to tie an economic value to changes in hunting quality. In the PPR, annual per-acre values of the effects of new wetlands on hunting quality ranged from near $0 to $143 per acre, with the highest values in the PPR's western fringe (figure 3).

ANALYTICAL METHODS FOR VALUING ENVIRONMENTAL AMENITIES

The following descriptions are meant to strengthen our illustration of the analytic framework supporting the valuation of wetland benefits. For more details on these and additional approaches and their application to wet- land benefit analyses, see Brander et al. (2006).

Contingent Valuation relies on surveys of a sample of the relevant population. Individuals are asked about their willingness to pay for (or accept) a change in an environmental amenity. There are many variations of this approach, but they involve surveys on preferences. The collected data are then used to estimate the public's demand (or willingness to pay) for the amenity (Loomis and Walsh, 1997).

Travel Cost Method uses survey data that included details on the quality of the amenity each survey respondent sees, personal information, travel time and costs, and recreation activities associated with the amenity. Changes in consumer surplus associated with a change in the amenity are derived from the estimated demand function.

Hedonic Analyses derive social values of environmental amenities by estimating demands for market goods as a function of (among other things) the quantity and quality of the amenities. This approach is suitable when the existence of an amenity directly affects the price individuals are willing to pay for a market good. For example, one can value an improvement in a lake's water quality by estimating consequential changes in the market prices of properties around the lake.

Damage Function Approach is based on the assumption that the value of a change in an amenity (such as the economic impact of a flood) is approximately equal to the cost of repair damages, loss in profits, and decrease in standards of living associated with the change in the amenity. The approach is thought to provide imprecise benefit estimates because it implicitly assumes that no remedial actions are taken and there is no accounting for market effects. For example, a reservoir can reduce flood damages. The value of the reservoir's flood protection benefit, based on the damage function approach, is the reduction in expenditures made to repair flood damages. But, with reduced risks of flooding, individuals will reduce expenditures on flood-damage prevention. This cost savings is also a flood protection benefit.

Replacement Cost Method assumes that the welfare effect of a loss in an amenity is approximately equal to any increase in expenditures that would subsequently restore the amenity. For example, the loss of safe drinking water from one's well is approximately equal to one's subsequent purchases of bottled water. This approach is made difficult because changes in expenditures may not be observed, and remedial action may not be taken.

Averting (Defensive) Expenditure Method assumes that the welfare effect of public actions taken to reduce or prevent impacts to an amenity (e.g., reduce downstream flood damages) is approximately equal to the dollar amount individuals pay to reduce or avoid the impacts (such as their cost of sandbagging floodwaters). Some studies argue that this approach overstates benefits (Freeman, 1993; Ribaudo, 1989). Although, in practice, it can be difficult to isolate the portion of expenditures that are attributable to averting damages, the approach continues to be used (Whitehead, 2005).

Benefits Transfer is a valuation approach used to generate benefit estimates based on the results of past research. There are (roughly) three types of benefit transfer approaches. The value transfer approach reviews the economic literature for per-unit values (or average values) that are then used as a measure of value. The second approach uses past benefit functions and their results to generate more generalized benefit or willingness-to-pay functions. And meta-analyses estimate benefit functions using findings of past research as independent variables. While this approach is convenient, it has imposed substantial errors (Brander et al., 2006).

The biophysical models: Duck populations

Duck population growth is constrained by the availability of breeding habitat and factors affecting nest success (Hoekman et al., 2002). In other words, an increase in wetlands that support duck reproduction will increase duck populations (a wetland service). We developed models that generate spatial estimates of expected increases in numbers of hatchlings (of the five dominant duck species) produced by new wetlands in the PPR. The models were constructed using a combination of earlier models, findings of past research, and available data. The models simultaneously predict the number of duck nesting pairs that are likely to take residence in a new wetland and the probability that a nest is successful (that is, the eggs hatch), given its location and the surrounding vegetative cover (for details, see appendix 3). To determine the subsequent change in duck populations, we applied findings of

Krapu et al. (2000) and Nichols and Hines (1983), which suggest that about 50 percent of all duck hatchlings reach adulthood.

The effects of new wetlands on hatchling populations were found to vary substantially across the PPR, with the greatest impacts in the Dakotas along the Missouri River. We found that the effect of a new wetland can range from 0 to over 55 hatchlings (27.5 adult ducks) per wetland acre.

Valuation of amenities: Duck hunting benefits

Multiple studies evaluated the economic value associated with duck hunting. But, only three generated estimates of consumer surplus that result from an improvement in the quality of duck hunting on a scale suitable to analyses of national or regional conservation policies and programs (table 2).[16] Each of these studies defines hunting quality (the amenity) as the number of ducks bagged. Across these studies, willingness-to-pay estimates for a marginal improvement in hunting quality range from $15 to $140 per duck (in 2011 dollars). These studies have limitations (see box "Estimating the Value of One More Duck"). Based on our assessment of this literature, we use a value toward the upper end of this range ($106) and discuss how estimates and their implications are affected by lower duck-value estimates.

Linking services to amenities: Hunting quality

With ducks bagged as a proxy measure of hunting quality, we used prior research and reasonable assumptions to link changes in duck populations to changes in ducks bagged. Based on mallard harvest rates of approximately 10-12 percent and a blue-wing teal harvest rate of approximately 2.5 percent (U.S. Fish and Wildlife Service, 2012), we applied what we believe to be a conservative harvest rate of 5 percent. In other words, each additional duck has a 1-in-20 chance of being bagged. Using an amenity value of $106 per bagged duck, the expected value of an additional duck is approximately $5.00, or $2.50 for each additional hatchling. Multiplying the hatchling value by a wetland's effect on hatchlings, we find that, within the PPR, the annual per-acre values of wetlands duck-hunting benefits range from near $0 to $143, with the highest values in the western fringe of the PPR. Both the value of bagging a duck and the harvest rate directly affect the per-acre value of a wetland. For example, if one were to apply a $26 per-duck value (which is about 25 percent of the value we applied), per-acre values would fall by 75 percent—from near $0 to $46 per acre. If one were to apply a 10-percent harvest rate, per-acre benefits would double. Applying other per-duck values and harvest rates is equally straightforward. In all cases, the pattern of relative values shown in figure 3 remains unchanged.

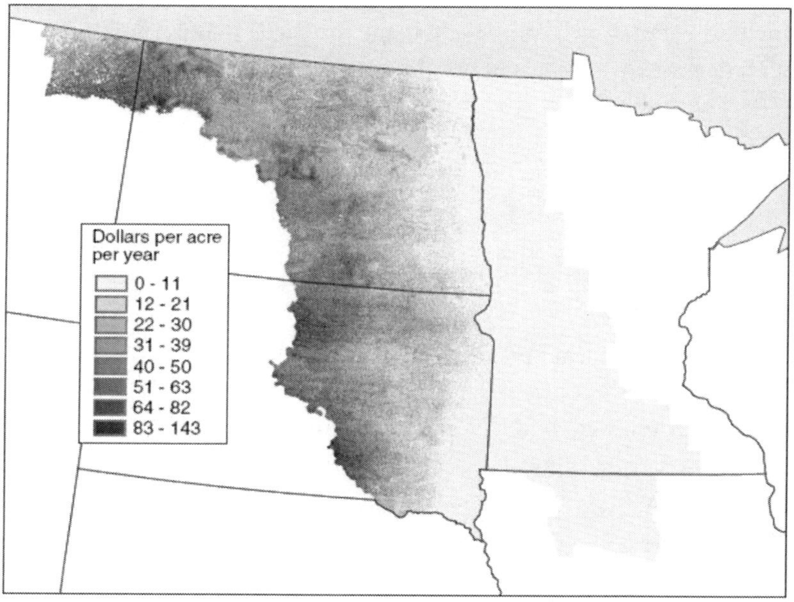

Source: USDA, Economic Research Service analysis of wetlands' effects on hunting quality (appendix 3) and a "per-duck" consumer surplus estimate of $106.

Figure 3. Value of new wetlands' effects on duck-hunting quality in the Prairie Pothole Region.

Table 2. Studies that estimate the change in consumer surplus associated with bagging an additional duck

Study	Valuation method and type of data	Consumer surplus per duck bagged
Hammack and Brown (1974)	Time series data on hunting activities	$15 to $34 with a mean value of $24.50
Charbonneau and Hay (1978)	Survey of hunting activities of hunters in the Mississippi River flyway	$106
Mackenzie (1993)	Survey of Delaware hunters on hypothetical hunting options	$132 to $140

Note: All values are reported in 2011 inflation-adjusted dollars. Source: USDA, Economic Research Service.

ESTIMATING THE VALUE OF ONE MORE DUCK

Our review of the literature revealed five core studies with objectives that include estimating hunters' demand for duck hunting.[1] Three studies specified demand as a function of (among other things) hunting quality (see table 2). These studies derive hunters' willingness to pay for a marginal change in duck hunting quality by differentiating the demand function with respect to hunting quality—bagging one more duck. The first (Hammack and Brown, 1974) used time-series data on hunting behavior in the Pacific Northwest. It estimated the gain in consumer surplus of bagging an additional duck to be between \$15.38 and \$33.62 (in 2011 inflation-adjusted dollars).

The second (Charbonneau and Hay, 1978) used cross-sectional data from the 1975 National Survey of Hunting, Fishing, and Wildlife Associated Recreation (NSHFWAR) to estimate demand as a function of hunting quality, experience, costs (travel expenses and the opportunity cost of travel time), and income. Based on their estimated demand function, the authors found that hunters are willing to pay, on average, \$106 to bag an additional duck on a single day (see table 2).

The third study (Mackenzie, 1993) used data from a survey of Delaware duck hunters where hunters were asked to choose from alternative hypothetical sites to estimate hunters' demand for duck and goose hunting. Similar to the approach of Charbonneau and Hay (1978), demand was estimated as a function of travel cost, travel time, the site characteristics, hunting partners, and site congestion. Mackenzie found that, on average, hunters' willingness to pay for an additional duck bagged ranged from \$132 to \$140 (see table 2).

Two other studies have directly estimated demands for duck hunting; neither estimated WTP for improved hunting quality. Each generated per-day consumer surplus estimates, as did Charbonneau and Hay (1978). We therefore compare the day-value estimates to gain some insight into the consistencies of the estimated demand functions. Charbonneau and Hay (1978) estimated per-day consumer surplus to be \$121. Brown and Hay (1987) —employing the 1980 NSHFWAR data—estimated per-day consumer surplus to range from \$64 to \$82, which is about 65 to 80 percent of the Charbonneau and Hay (1978) estimate.

However, Brown et al. (1978) — using the same data as Charbonneau and Hay—estimated average per-day consumer surplus to be $122, which supports the findings of Charbonneau and Hay (1978).

We chose not to use Hammack and Brown (1974) because, unlike subsequent research, they failed to account for the opportunity cost of time—the opportunity cost continues to be recognized as a critical part of travel cost (Larson and Lew, 2014; Menon et al., 2014). Additionally, Gascoigne et al. (2011) rejected the Hammack and Brown (1974) model because, upon the use of more recent values of the independent variables, they felt the model generated values that were unacceptably low (less than $4 per duck).[2]

We applied the Charbonneau and Hay estimate of $106. The more recent Mackenzie (1993) results suggest this value might be low. Conversely, our day-value comparison with the Brown and Hay (1987) estimates suggests the Charbonneau and Hay (1978) estimate might be high.

The age of these studies raises questions: Has the public's willingness to pay changed over time? Has the increase in the U.S. population increased the number of duck hunters and the number of people benefiting from improved hunting quality? Have behavioral models or empirical methods improved in such a way that duck hunting demand might be better estimated? These same questions arise in most meta analyses. As with authors of previous studies, we assume that, after adjusting for price-level changes, the reported values are reasonable. But given our uncertainty of this estimate, we also use other duck values to illustrate the sensitivity of our results.

[1]These five studies estimate hunters' demand for hunting across wide regions—they estimate the aggregate demand within their study area. Many other studies estimate hunters' demand at a single site or across substitute sites, such as game refuges in California. Demand at each site depends on the size and interest of the surrounding population, travel costs, the quality of hunting at surrounding sites, and other location-specific factors. If the results of a substantial number of site-demand studies scattered throughout the relevant region were available, it may be possible to estimate an aggregate demand based on the results of the site-level studies. However, instead of the studies being scattered throughout, the majority of the studies focus on sites on public lands in areas where hunting is popular.

[2]Gascoigne et al. (2011) subsequently moved from a marginal benefit analysis to an aggregate day-value approach.

Nitrogen removal: What the study found

Wetlands can improve water quality by removing chemicals before they reach streams, rivers, and other waterways. Of particular interest is the ability of wetlands to filter out reactive nitrogen (N).[17] Reducing nitrogen loadings in water provides various amenities:

- Improvements in the quality of municipal intake waters
- Cleaner water for swimming, boating, and recreational fishing
- Increased profitability of commercial fisheries

Confined by the limitations of data and models, we evaluated N removal by new wetlands in the Upper Mississippi and Ohio River watersheds—this area is the primary source of N affecting the hypoxic zone of the Gulf of Mexico, where excess nutrients have depleted oxygen needed to support marine life. Additionally, this region has opportunities for wetland restoration because 80 to 90 percent of prior-existing wetlands were drained. But we could not link the change in nitrogen loadings to an impact on an environmental amenity, nor could we find or estimate a monetary value for impacts on amenities. We therefore focused on N removal and found that rates varied substantially, ranging from 11 pounds to nearly 1 ton per wetland acre per year (see figure 4).

The biophysical model: Nitrogen removal

Building on available models and data, we generated estimates of the expected quantity of N removed by a restored 10-acre wetland. These models work with hydrology, inflow water quality, and weather and climatic conditions at grid points throughout much of the Upper Mississippi and Ohio River watersheds.[18] To capture the effects of daily and annual variations in precipitation levels and intensity on nitrogen movement and wetland effectiveness, we ran 20-year simulations (see appendix 4).

Economic models of amenities affected by nitrogen inflows

We found two large-scale economic analyses of the benefits of reducing N loadings; both relied on reduced-form models. First, in an attempt to value changes in Gulf of Mexico hypoxia, Diaz and Solow (1999) used 30-year time-series and geographical data to estimate economic impacts on commercial fisheries and recreational fishing activities as a function of the size of the hypoxic zone and other factors and found no statistically significant results.

Second, in an analysis of municipal water treatment costs, Ribaudo et al. (2011) found that nitrogen has a statistically significant impact on costs. They reported municipalities' longrun nitrogen removal costs ranging from $20 to $36 per million gallons of treated water in those areas where treatment plants are necessary. Note that this cost estimate is a function of the volume of water and not the quantity of nitrogen. Even so, the study's findings suggest that a change in nitrogen inflow to streams affects water treatment costs.

At the national level, the U.S. Environmental Protection Agency Science Advisory Board (2007) examined both the benefits and co-benefits of nitrogen conservation practices, including those at wetlands. The Board found evidence of economic benefits of N reductions but concluded that co-benefits (such as the values of consequential decreases in soil erosion) may exceed the nitrogen reduction benefits.

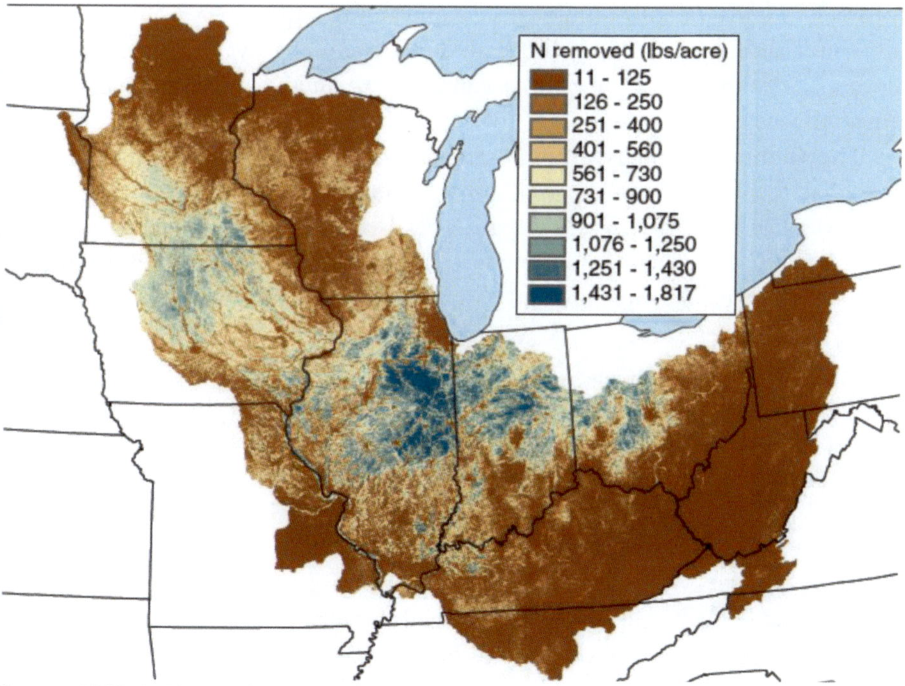

Source: USDA, Economic Research Service.

Figure 4. Amount of nitrogen removed annually by a new wetland in the Upper Mississippi and Ohio river watersheds.

Linking services to amenities: The value of reduced n inflows

Overall, we were unable to link spatial changes in N flows into streams to the subsequent outflows to the Gulf of Mexico. But Rabotyagov et al. (2014) found a statistically significant link between N inflow to the Gulf of Mexico and the size of the hypoxic zone, using a simple statistical model and time-series data.

Overall, evidence of the effects of the Gulf of Mexico's hypoxic zone on ecosystems is mixed. Bianchi et al. (2010) found no evidence that discharge from the Mississippi River adversely impacted Gulf of Mexico fisheries. Consistent with the findings of Diaz and Solow (1999), other studies found no clear link between hypoxia and changes in fishery landings in the Gulf (Rabalais and Turner, 2001). In fact, Bianchi et al. (2010) note that there appears to be some effect on the distributions of mobile fish species and phyto-and zooplankton that prosper in the nutrient-rich waters that add to, what appears to be, an additional food web.

Still, there are grounds for concern about hypoxia in the Gulf of Mexico. For example, the Federal- State Mississippi River Gulf of Mexico Watershed Nutrient Task Force (2013) noted that "low dissolved oxygen in the Gulf is a serious environmental concern that can impact valuable fisheries and disrupt sensitive ecosystems."

The findings of analyses of waters in the Gulf of Mexico must not be viewed as representative of other hypoxic zones. The hypoxic zone of the Gulf of Mexico lies within a great expanse of deep waters while, in most other cases, hypoxic zones are in shallow waters in bays and along shorelines where ecosystems are easily impaired and damaged (Bianchi et al., 2010).

Species Protection: What the Study Found

The public's willingness to pay to protect species reflects the following:

- A desire for opportunities to view the species
- The ethics of protecting resources for future generations
- Uncertainty that there may be subsequent domino effects on other species if one or more species or their habitats are lost

This analysis considers only wetland-associated species identified by NatureServe as imperiled in 2014.[19] In 2014, approximately 597 wetland-associated species were listed as imperiled (217 are listed as critically imperiled) across the entire range for each in the United States.20 The number of wetland-associated imperiled species within a county ranges from 0 to 23 (figure 5a).

The need for a biophysical model: Species protection

An appropriate biophysical model would quantitatively link the addition of a wetland to the affected species. We obtained data from NatureServe (2014) that lists each wetland-associated imperiled species and the counties where each exists. The data are insightful but do not indicate whether a species' survival is affected by an additional wetland—other habitat conditions, such as water quality, predation, and migratory habitat, may have an effect. Even in counties with no wetland- associated imperiled species, it is possible that an additional wetland could have a positive impact on imperiled species—that is, expanding wetland habitat could help imperiled species in neighboring or downstream counties.

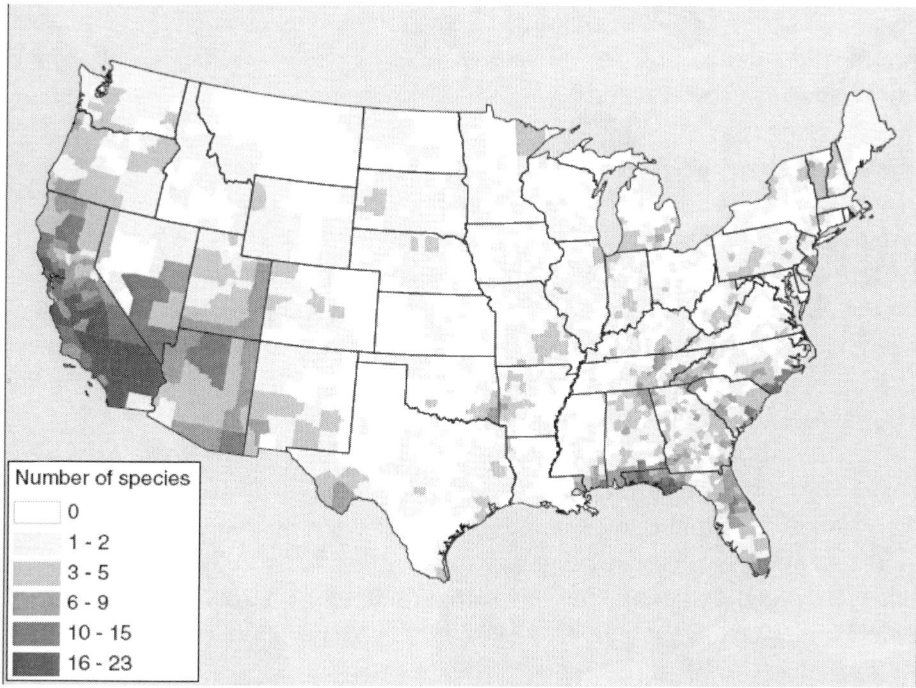

Source: USDA, Economic Research Service using data from NatureServe, 2014.

Figure 5a. Wetland-associated imperiled species, by county.

Valuation of amenities: Species protection

Some view species protection as a moral or ethical issue and object to placing a value on protecting a species. However, Shogren et al. (1998) point out that, while most Americans believe in fully protecting endangered species, they are unwilling to bear the cost of increasing the strength of the 1973 Endangered Species Act. Because public funding for species protection is limited, decision- makers who prioritize wetland conservation efforts consider tradeoffs in species' protection so that, implicitly or explicitly, species are being valued (Roughgarden, 1995).

Past research suggests that the public tends to have a higher willingness to pay to protect so-called "charismatic" species, including many large birds and mammals, than to protect others (Loomis and White, 1996; Brander et al., 2006; Richardson and Loomis, 2009; and Woodward and Wui, 2001). In developing data for their meta analysis, Richardson and Loomis (2009) collected published species valuation studies. Most of the studies they found used contingent valuation, where respondents are asked a series of questions that are designed to reveal their willingness to pay to protect one or more species. However, the reported values are often based on a generalized concept of protection that cannot be directly used to calculate values of marginal improvements in species' survival.

Amenities provided: Wetland effects on imperiled species

An increase in a species' survival probability is a quantitative measure of an increase in an environmental amenity (see 5-8 in U.S. Environmental Protection Agency (2014)). Research has yet to develop biophysical models that quantify new wetlands' effects on species survival probabilities (e.g., the amenities). Without a measure of new wetlands effects, estimates of the values of changes in species' survival probabilities offer little insight into the relative importance of different wetlands' effect on imperiled species.

In the absence of biophysical models, we have generated proxies to provide insight into the effects of wetland restoration on imperiled species. The proxy is the number of wetland-associated imperiled species in a county. This approach has been used in past studies—counties with high imperiled or endangered species counts are considered "hot spots" (Dobson et al., 1997; Ando et al., 1998). Szentandrasi et al. (1995) used this approach to assess how one might target CRP enrollments.

An analysis of NatureServe data shows that imperiled species were located in 1,745 counties (about 57 percent of all contiguous U.S. counties) in

2014, and that 1,366 counties have no wetland-associated imperiled species. Concentrations of imperiled species are greatest in the Southwest (as many as 23 per county) and are spread throughout much of the Southeast (figure 5a).

Restricting the analysis to vertebrate species (such as birds, amphibians, and mammals), which tend to receive the most public attention, we found a total of 76 imperiled wetland-associated species (28 of which are critically imperiled) and numbers per county ranging from 0 (in 2,010 counties) to 8. Southern California and neighboring lands and Appalachia are the hot spots (figure 5b).[21]

These findings must be cautiously interpreted because there are no measures of the effects of a new wetland or the values of effects on different species. Findings do, however, provide some perspective on where new wetlands might benefit a greater number of imperiled species.

Flood Protection: What the Study Found

Society is willing to pay to reduce flooding because of private and public benefits including damages:

- To reduce personal risks of damage to property and health and a disruption of daily lives
- To reduce public expenditures on disaster relief
- To avoid damage to public infrastructure
- To maintain healthy communities
- To adhere to ethics of protecting the health and property of others

Wetlands provide protection by storing water, slowing its movement, and reducing storm surges associated with heavy precipitation and rapid snow melts. Losses caused by floods include both human life and property. Floods damage private structures and public infrastructures, destroy crops, and wash away farmland. According to the U.S. Department of Commerce, National Oceanic and Atmospheric Administration (NOAA), annual property damages attributed to flooding usually lie between $1 billion and $10 billion, but a spike of $50 billion occurred in 2005 due to a large number of storms and the especially severe damage wrought by Hurricane Katrina.

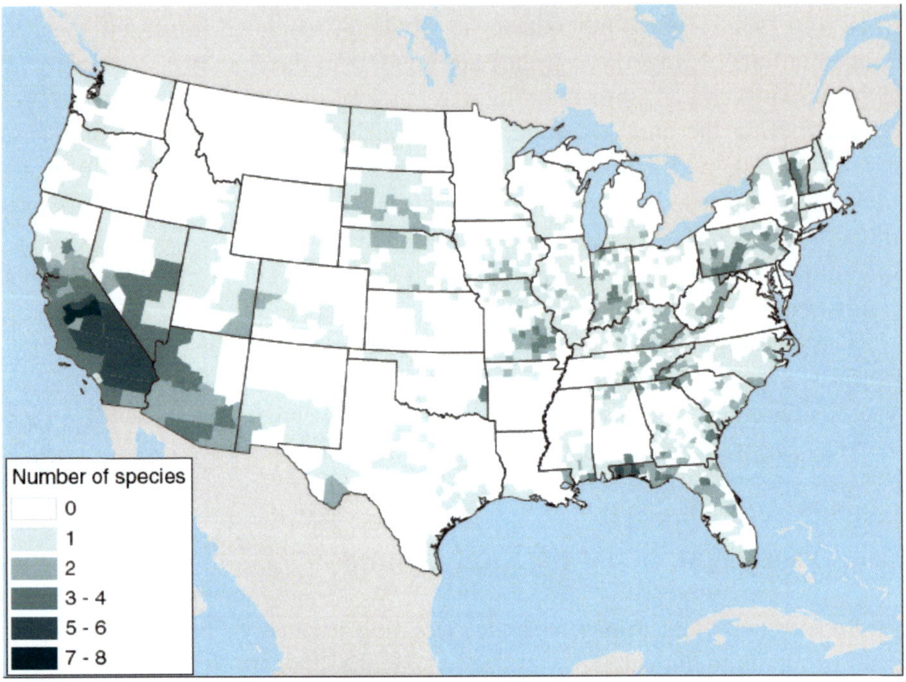

* Includes birds, mammals, amphibians, and reptiles.
Source: USDA, Economic Research Service using data from NatureServe, 2014.

Figure 5b. Wetland-associated imperiled vertebrae species,* by county.

The biophysical model: Floodwater storage

In a comprehensive analysis, Bullock and Acreman (2003) reviewed 169 studies of wetlands' hydrologic performances, including their effects on flooding. Findings of these studies varied widely—in some cases, wetlands increased flooding. They conclude that, while wetlands play a role in flood protection, comparisons of wetlands' effectiveness must be given careful consideration.

A wetland's "bounce"—its ability to store additional water as rivers reach peak flows—can help determine whether the wetland can control flooding. A wetland's bounce is usually small (Bullock and Acreman, 2003). However, bounce can be increased if water level can be controlled. For example, in a site-level analysis, Bengtson and Padmanabhan (1999) used a hydrologic model to estimate the relationship between reduced peak flow and storage bounce in the Red River Watershed of North Dakota. They found that, unless modified, wetlands in the watershed did little to reduce peak flows because, as river levels rose, wetlands filled. But, if wetlands were allowed to drain until rivers began to peak, a 1-foot impediment placed on all wetlands (creating a 1-foot storage bounce across the wetland acres) could reduce damages from high-frequency flood levels by 3.8 percent; a 2-foot impediment could reduce damages by over 5 percent.[22]

Valuation of amenities: Expected reductions in flood damages

Five studies in the literature estimated the economic value of wetland flood protection benefits. The studies do not offer values directly transferable to wetlands restored on agricultural lands for one or more of the following reasons: they focus on wetlands in close urban proximity with limited agricultural lands; they focus on coastal areas where agricultural acreage is limited; they evaluate wetlands with flow impediments; and they aggregate wetlands with other land uses. However, the studies are likely to illustrate upper-bound ranges of flood protection values.

Gupta and Foster (1975) estimated the flood control benefits of wetlands in the Charles River Basin of metropolitan Boston—a major urban area. Using Army Corps of Engineers data, they found expected avoided damages to be $56 to $440 per wetland acre (inflation adjusted to 2011 dollars) per year. Values such as these may be applicable to wetlands on agricultural lands but in limited areas with considerable housing or urban development.

NOAA provides statistics on some U.S. coastal wetlands' services, including estimates of coastal wetlands' flood protection benefits that were originally generated by Costanza et al. (2008) (see http://stateofthecoast.noaa.gov/wetlands/welcome.html). Costanza et al. reported benefits of near $0 to $29,600 per acre per year, with some of the highest benefits in Louisiana. Conversely, Farber (1987) estimated benefits of flood protection by coastal wetlands in Louisiana to be substantially less (i.e., less than $50 per acre). Again, values such as these may be applicable to wetlands on agricultural lands in coastal areas.

Shultz and Leitch (2003) estimated the flood protection benefits of restoring 2,700 acres of wetlands—with flow impediments—in the Maple and Red River Watersheds of North Dakota. They found that when wetland outlets were controlled (e.g., water was captured only when flows were peaking, consistent with Bengtson and Padmanabhan (1999)), expected annual benefits ranged from $12 to $25 per acre.

The U.S. Army Corps of Engineers (2013) estimated flood damage reduction benefits to downstream urban areas due to the CRP's wetlands and riparian buffers. This pilot study focused on a small Iowa watershed (Indian

Creek). It used several hydrologic models, simulated weather events, and a variety of CRP coverage scenarios (where each scenario differed in the quantity of CRP acreage and the distributions between wetlands and riparian buffers). The study found that, when rainfall events were short but very heavy, acres under CRP wetland practices can reduce property damage by 11 percent (although acres in riparian buffers are more cost effective). When the methods of this study were applied to a larger Iowa watershed (Cedar River), there was no measureable relationship between the CRP's wetland and riparian buffer acreages and the change in flood stage.

Amenities provided: Reduced flooding due to wetlands

The dollar-per-acre flood protection benefits reported in Shultz and Leitch (2003) and other studies embody the link between the effects of wetlands on flooding and flood damages—they are reduced- form analyses in which the benefit estimates are dependent on physical and economic parameters. We found too few studies in the literature that generated spatial estimates of wetlands' flood protection benefits within any region to enable us to develop benefit models through benefits-transfer analyses.

Net Greenhouse Gas Emissions: What the Study Found

Many countries are seeking to reduce the growth in atmospheric concentration of greenhouse gases (GHG), primarily by reducing emissions. Some countries are also engaged in efforts to remove GHGs from the atmosphere, commonly through carbon sequestration.

Wetlands are known to reduce atmospheric concentrations of carbon dioxide through carbon sequestration. But wetlands are also known to emit GHGs, primarily methane (CH_4) and nitrous oxide (N_2O). Both of these are considered "hot" and can more than offset the effect of CO_2 removal.

Past research that has reviewed and evaluated findings of studies worldwide (Kayranli et al., 2010), within North America (Bridgham et al., 2006), and within the tropics and subtropics (Mitsch et al., 2012) concluded that the effects of wetlands on GHG emissions are questionable, as carbon sequestration may be more than offset by methane and nitrous oxide emissions. These studies illustrate the importance of accounting for net GHG effects, but they do not provide measures of the GHG effects of different types and ages of wetlands. Similarly, many studies have estimated carbon sequestration and carbon storage by different types of wetlands (for an overview, see Bernal and Mitsch, 2012), but most do not measure net GHG impacts.

Euliss et al. (2006) and U.S. Fish and Wildlife Service (2005) evaluated the net GHG effects of wetlands in the PPR and Mississippi Alluvial Valley (MAV) (see figure 1), and findings from these studies are used in this analysis along with estimates of the marginal social cost of carbon developed by a Presidential Commission (U.S. Government, 2010). Based on these data, the annualized carbon sequestration benefit of a new, permanently restored wetland is $55 per acre in the PPR and $129 for an MAV wetland (table 3).

Net greenhouse gas emissions: The biophysical models—GHG impacts

Past research provides estimates of the net GHG impacts of wetlands in two regions: the PPR and MAV. Euliss et al. (2006) estimated that, 5 years after a wetland in the PPR is restored, 7.8 metric tons per acre of carbon is held in the living biota above and below ground—an average sequestration rate of 1.56 metric tons per acre per year. In subsequent years, carbon sequestration was estimated to average 0.34 metric tons per acre per year. Additionally, while wetlands of the PPR were estimated to release methane and nitrous oxide, the release rates were estimated to equal the releases on the lands' prior wetland restoration, meaning that the restored wetlands led to no change in the releases of these gases (Gleason et al., 2009; Gleason et al., 2005; Eagle et al., 2012).

Table 3. Annual average per-acre carbon sequestration benefits of Prairie Pothole Region and Mississippi Alluvial Valley wetlands

Wetland regions*	SRD=3 r=4	SRD=5 r=4	SRD=2.5 r=4	SRD=3 r=5	SRD=3 r=3
			*$/acre**		
PPR	1,378	371	2,160	1,251	1,536
MAV	3,222	930	4,955	2,776	3,667
Annualized values		-----------------4%-----------------		5%	3%
PPR	55	15	86	62	46
MAV	129	37	198	139	110

PPR=prairie pothole region; MAV=Mississippi Alluvial Valley: SRD=social rate of discount (%); r=market discount rate (%).

*2011 dollars.

Source: USDA, Economic Research Service.

In a second study in the PPR, Gleason et al. (2008) estimated carbon in above- and below-ground vegetation after 10 years to be about 11.13 metric tons per acre, which is about 1.6 metric tons per acre greater than the amount estimated by Euliss et al. (2006). We applied the Euliss et al. estimates because they are more conservative and they account for the effect of time on sequestration rates.

Within the MAV, carbon accumulates primarily in the vegetative biomass (Faulkner et al., 2008). In 2005, the U.S. Fish and Wildlife Service (FWS) reported that vegetation on newly restored MAV wetlands held 99 metric tons per acre of carbon after 70 years for an average annual rate of 1.41 metric tons per acre per year (USFWS, 2005). A second FWS analysis, also released in 2005, reports a 70-year average carbon sequestration rate of 1.77 metric tons per acre per year. We applied the more conservative 1.41 metric tons per acre per year. Research on N_2O emissions by restored forested wetlands found no difference from the level released on agricultural cropland (in the first 70 years), while net effects on CH_4 releases are unknown but are not expected to have a large effect (Faulkner et al., 2008).

The GHG impacts reported here are only relevant in cases where cropland is converted to wetlands. As with all evaluations of the GHG impacts of conservation practices, the prior use of the land and its effect on carbon loadings and releases of GHGs must be considered as it defines baseline conditions (see box "The Value of GHG Impacts Depends on GHG Emissions Over Time"). For example, pasture and hay land usually hold a greater stock of carbon in above- and below-ground vegetation than cropland. Therefore, the GHG impacts of wetlands restored on pasture or hay land will likely be lower than those of wetlands restored on cropland.

Valuation of amenities: Net greenhouse gas emissions

To meet Executive Order 12866, an interagency research team was assembled to generate estimates of the social cost of carbon (SCC).[23] These estimates are measures of the marginal cost imposed by a 1-ton change in atmospheric carbon dioxide in an observation year. Because the marginal economic effects increase as GHGs accumulate, the SCC increases over time (see table A1 in U.S. Government, 2010).[24] The Executive Order requires agencies to use the estimates in cost-benefit analyses of policies and actions that have marginal impacts on GHGs (for details, see U.S. Government, 2010).

The SCC estimates—calculated using the damage function approach—are based on agreed-upon values related to many controversial issues, including the value of a human life, the expected rate of GHG accumulation, technological advances that offset impacts, and—perhaps the most economically controversial—the rate at which future damages (or adverse impacts to individuals' life and standard of living) are discounted, or social rate of discount (SRD) (U.S. Government, 2010). Some observers believe that the SRD should be 0 so that, for example, the value of a lost life in the future is the same as a lost life today; others believe a market discount rate is appropriate. This study applied SCC estimates that were derived from the agreed-upon SRD of 3 percent. For more on the conceptual differences between market and social discount rates and the sensitivity of the SCC estimates to the SRD, see U.S. Government (2010).

Amenities provided: Wetland impacts on GHG

Each year, the values of a new wetland's GHG-reduction benefits differ due to variations in GHG impacts and the SCC. Instead of listing annual values, this study capitalized them using a 4-percent market rate of discount.[25] We found that the present value of the GHG impacts of a restored wetland in the PPR is $1,378 per acre and $3,222 for an MAV wetland (assuming the wetland was restored in 2013). Our estimates assume there will be no changes in wetlands' GHG benefits after 2050 because there are no SCC estimates beyond this date. Had there been estimates for more years, our estimates of the GHG benefit of the restored and then undisturbed wetlands of the PPR and MAV would probably have been greater. To compare GHG benefits with our estimates of annual benefits and costs, we annualized total benefits across the wetlands' permanent lives using a 4-percent discount rate (as used in our cost analysis) and found regional values of $55 and $129 per acre, respectively (see table 3).

These benefit estimates are based on *permanent* restoration. If a wetland is moved back into agricultural production, carbon is released and GHG impacts become negative. For example, the present value of the carbon sequestration benefits of a permanently restored PPR wetland can be 20 times the value of a wetland that is converted to agriculture after 10 years (see box "The Value of GHG Impacts Depends on GHG Emissions Over Time"). All economic analyses of GHG benefits of conservation practices need to account for longevity. Conversely, losses of wetlands eliminate the flow of annual benefits provided, such as improved water quality and better hunting quality, but there are no subsequent damages (e.g., no negative costs imposed).

THE VALUE OF GHG IMPACTS DEPENDS ON GHG EMISSIONS OVER TIME

The carbon cycle associated with an agricultural practice is the annual change in atmospheric greenhouse gases (GHGs) over time, given what they would have been. This means that one must consider lands' GHGs effects with and without the practice, even after the practice is terminated.

Each social cost of carbon (SCC) estimate applied here is defined as the present-value sum of future damages of the climate change impact due to a marginal change in atmospheric GHG concentrations (see U.S. Government, 2010). The SCC estimates increase over time because, as GHG concentrations increase, marginal damages increase.

Because the SCC rises over time (though the present value of future SCC estimates fall), it is critical to account for changes in GHG loadings on a year-by-year basis. This means one must account for effects after a conservation project's life. By doing so, one finds that permanent conservation practices can generate far more benefits. For example, if a wetland restored (or tillage practice begun) at the beginning of 2013 sequesters 1 ton of carbon per year, its GHG benefits (assuming no impacts on other GHGs) would be $24.71 in 2013, $25.28 in 2014, etc.[1] If the wetland is returned to agricultural production in 2015 and releases 1 ton of carbon each year, the damage due to the ton released in 2015 would be $25.82 and $26.00 for the ton released in 2016. The net GHG benefit of the wetland would be $2.09, when discounted to 2013 ($2.09 =24.74 +25.28*(1.04)-1 – 25.82*(1.04)-2 – 26.37*(1.04)-3). Should the same conditions hold and the wetland have a 10-year life, the net present value of the GHG impacts would be $34.00. Finally, if the wetland was left undisturbed, the net present value of the GHG impact would be $623—nearly 20 times the value of the 10-year life and 300 times the value of the 2-year wetland.

[1] Values are from U.S. Government (2010) in 2011 dollars.

Groundwater Recharge, Sediment Removal, and Open Space: What the Study Found

No regional studies or data measure or value the effects of wetland restoration on groundwater recharge, sediment removal, or open space. Based on a number of small-scale studies, we concluded that (1) wetlands will sometimes provide positive recharge, sediment removal, and open-space benefits, but the sediment-trapping and open-space benefits can be negative; and (2) the benefits for each must be considered on a case-by-case basis.

Groundwater recharge: The analyses

The available literature indicates that, in some locations, wetlands may increase amenities through groundwater recharge (that is, the flow of water to groundwater supplies). But currently, little information is available on the effect of wetlands on groundwater.

In most cases, wetlands do not recharge groundwater supplies. Some wetlands are wet because they are fed by groundwater, and some have a mutual exchange of water such that wetlands hold water when water tables are high and release water as the water table drops. In other cases, the soils under the wetlands are impermeable (Smith et al., 2011) (for more details on wetland types, see www. water.ncsu.edu/watershedss/info/wetlands/function.html). However, wetlands in floodplains are often permeable and provide recharge that can help maintain flows in dry conditions (Bullock and Acreman, 2003).

The value of water added to groundwater supplies depends on how the water might serve the public. Should the groundwater be a source of drinking or irrigation water, the affected individuals will value reduced pumping costs and any lessening of water-supply constraints. People will, indirectly, value groundwater that feeds lakes and streams if it subsequently increases amenities, such as fishing quality, species protection, and scenic beauty.

Sediment removal: The analyses

Based on the available literature, we were unable to conclude whether sediment trapping by any type of wetland will have net positive benefits. By trapping sediment, wetlands can improve water quality, but the sediment can damage the wetland ecosystems and, over time, fill some types of wetlands and eliminate subsequent wetland benefits (Gleason and Euliss, 1998). Other types of wetlands, such as riparian wetlands, may periodically receive high water flows that flush sediment into deeper waters and waterways; the flushed sediment can have adverse environmental impacts.

Open space: The analyses

We were unable to find studies that report the value of open space in rural settings. This is not unexpected because in rural areas, the public benefits of an additional acre of generic "open space" are likely to be small.[26] However, wetlands—as natural open spaces—may have open space values greater than lands without wetlands. But a few factors may reduce the size of, or offset, this extra value. First, since these wetlands are on private land, they may not be accessible (or visible) to the public, which may reduce or eliminate open-space values. Second, "slippage" may occur. Specifically, a new wetland easement will preserve the open space of land in the easement, but it will not prevent the loss of open space anywhere else. The value of the open space that is preserved by the easement can be offset when land is subsequently developed elsewhere (Lichtenberg, 2011; Irwin and Bockstael, 2004).

Targeting Wetland Conservation Efforts

In this section, we overlay the spatial estimates of annual wetland benefits and costs and highlight implications for targeting wetland conservation funding to maximize these benefits relative to costs.[27]

Duck Hunting Benefits and Costs of Prairie Pothole Wetlands

Duck-hunting benefit-cost ratios are greater than one in about one-third of the PPR and are greatest in Montana and parts of western North Dakota, driven by both lower land costs and higher ecosystem productivity (figure 6). This finding suggests that, in general, wetland restoration funds spent in the western PPR are likely to be a good investment in that they are likely to provide a positive social rate of return. Note that we have not conducted a site-by-site analysis, so some specific sites might have benefit-cost ratios that may or may not exceed our area estimate.

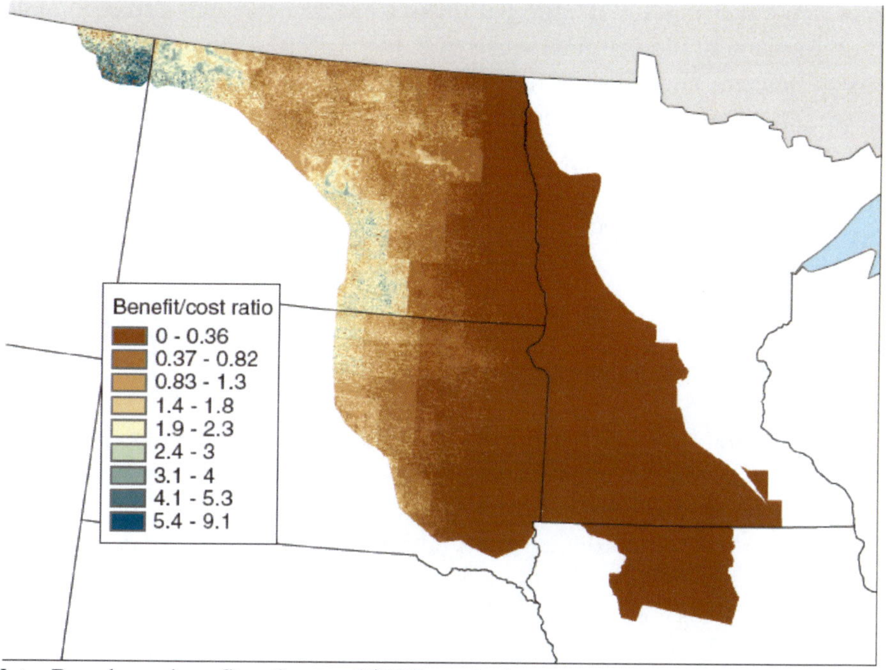

Note: Based on a benefit estimate of $106 per duck.

Source: USDA, Economic Research Service analysis of the value of wetlands' duck hunting benefits in figure 3 and wetland costs in figure 1.

Figure 6. Duck-hunting benefits relative to wetland cost in the Prairie Pothole Region.

Factors to consider when applying the duck-hunting benefit-cost estimates:

- If the value of new hatchlings differs across flyways, our benefit estimates may be missing some variation in wetland values.
- Other values of changes in duck populations, such as the value to birdwatchers, were not included because we found no measures.
- Wetlands of the PPR provide important duck nesting habitat (50 to 80 percent of all North American ducks nest in the PPR (Batt et al., 1989)). Wetlands in other parts of the country, suitable to nesting, will also increase duck populations and improve hunting quality.
- Because duck populations are constrained by wetlands that can support nesting populations (Reynolds et al., 2007), improving or expanding wetlands in areas where ducks do not nest is not expected to increase duck populations.

- We use a value of $106 per duck. Changing this value (for example, assuming the marginal value of a duck is $26) will cut benefits by 75 percent and reduce the size but not the spatial pattern of the distribution of the benefit-cost ratios. Conversely, if bag rates are greater than 4 percent (a value we feel is conservative), per-acre benefits would be greater. For example, increasing the bag rates to 10 percent would double the benefits amount.

Nitrogen Removal: The Cost Effectiveness of Wetlands of the Upper Mississippi and Ohio River Watersheds

The amount of nitrogen removed per dollar spent to restore and preserve a new wetland ranges from 0.15 to 34 pounds (figure 7). These estimates indicate that nitrogen removal costs, via new wetlands, range from $0.03 to $7.00 per pound. Within the area of study (the Upper Mississippi/Ohio River watershed), wetland restoration in east-central Illinois and southwestern Ohio tends to be most cost effective. If nitrogen reduction was the only environmental goal, these areas would be a good place to target wetland restoration funding.

Variations in costs and nitrogen removal rates both are strong drivers of cost-effectiveness. For example, east-central Illinois, southwestern Ohio, and north-central Iowa have some of the highest wetland nitrogen removal rates in the region (see figure 4). However, high wetland costs in north-central Iowa make wetland restoration in this area less cost effective than in the other areas mentioned (see figure 7).

Although nitrogen removal is not the only way that wetlands improve water quality, factors to consider when applying the nitrogen removal cost-effectiveness estimates include the following:

- The social values of reduced nitrogen loadings are unknown. But it appears that the presence of the hypoxic zone, and the evidence that schools of fish avoid it, increases the costs of harvesting fish due to an increase in travel cost. Also, public funding aimed at reducing nitrogen inflows to the Gulf of Mexico suggests that wetland impacts on nitrogen are likely to be of policy interest.
- There is little similarity between hypoxia's effect on the ecosystems of the open and deep waters of the Gulf of Mexico and the effect on the much shallower waters of bays and along shorelines.

- The cost-effectiveness estimates reflect reduction in nitrogen entering a stream, lake, or river. The estimates do not necessarily reflect the cost effectiveness of reducing downstream impacts because they do not account for in-stream nitrification.
- The costs and the effectiveness of wetlands vary by wetland size. Because of limits to modeling capabilities, this analysis considers only 10-acre wetlands.

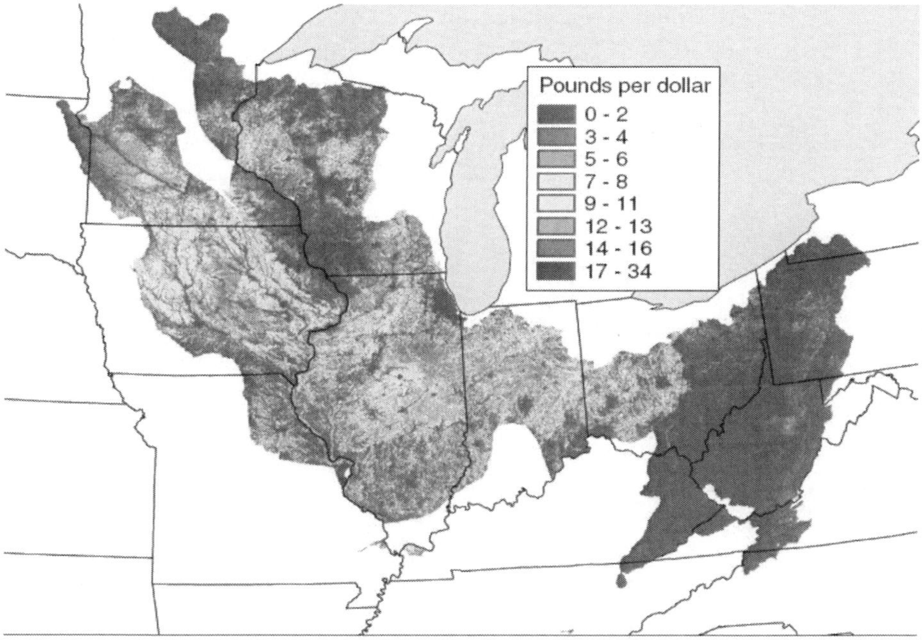

Source: USDA, Economic Research Service analysis of the estimated quantity of nitrogen that a new wetland can remove (figure 4) and the cost of new wetlands (figure 1).

Figure 7. Quantity of nitrogen removed relative to wetland cost in the Upper Mississippi and Ohio River watersheds.

Survival of Imperiled Species

Research results do not reveal how a change in wetland acreage will affect species' survival probabilities. Also, with only a few estimates of people's willingness to pay for a marginal increase in survival probabilities, we cannot value wetlands' species protection benefits.

We did examine "hot spots"—areas with large numbers of wetland-associated imperiled species. Dividing by cost, we found species-per-dollar estimates to be greatest in much of California, in southern Oregon, along the Missouri River in the PPR, and in Appalachia (figures 8a and 8b).

Factors to consider when applying the cost-effectiveness estimates:

- We identify counties with some of the highest numbers of wetland-associated imperiled species and report ratios of the numbers of species relative to wetland costs. These numbers provide a perspec tive but in no way indicate economic advantages of restoring one wetland relative to another.

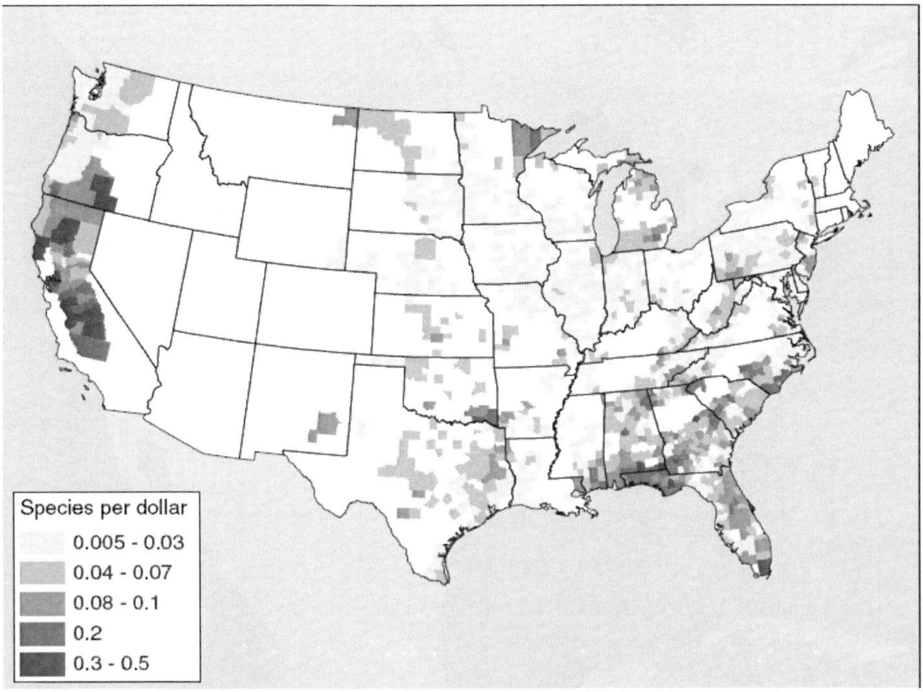

*A classification of "imperiled" is based on various biological and external factors that may affect the long-term viability of a species (Regan et al., 2004).

Note: White indicates that there are either no imperiled species or no cost estimates.

Source: USDA, Economic Research Service using data from NatureServe (2014) and the cost of new wetlands in figure 1.

Figure 8a. Number of wetland-associated imperiled* species relative to the cost of new wetlands, by county.

- The available data identify counties where no or very few wetland-associated imperiled species currently exist, which suggests, but does not indicate, that species-protection benefits might be relatively low.

Carbon Sequestration in the PPR and MAV

The value of the GHG impacts of wetlands throughout the MAV and much of the central and western PPR, the two areas where wetlands' GHG impacts have been estimated, exceeds our estimate of the cost of restoring and protecting wetlands (figure 9). If a 5-percent SRD was applied, wetlands' carbon sequestration benefit-cost ratios would fall by 70 percent. Even under this conservative scenario, benefit-cost ratios are greater than one in the northwestern part of the PPR and throughout much of the MAV.

Factors to consider when applying the carbon sequestration benefit-cost estimates:

- Benefit estimates assume that the wetlands are restored on agricultural land.
- Benefit estimates assume that the wetlands are permanent. Converting these wetlands back to agricultural use would result in the release of CO_2 and possibly a very substantial fall in benefits.
- The economic benefits of marginal changes in GHGs are those prescribed by Executive Order 12866 (U.S. Government, 2010). Agencies are required to use these values to assess the cost and benefits of GHG policy and program options.

Considering Benefits Together

Two of our benefit estimates overlap in the PPR: duck hunting and carbon sequestration. Summing the two PPR benefit-cost ratios, we find that wetlands in the western areas of the PPR have the highest ratios (figure 10). This is expected given the symmetry in the hunting and GHG cost-benefit ratios.

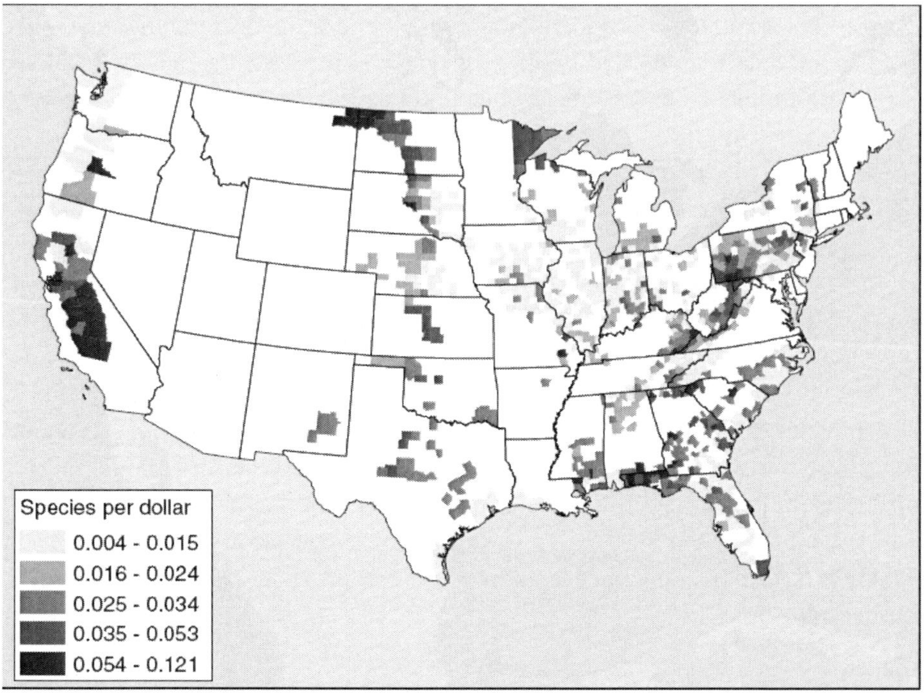

*Includes birds, mammals, amphibians, fish, and reptiles. **A classification of "imperiled" is based on various biological and external factors that may affect the long-term viability of a species (Regan et al., 2004).

Note: White indicates that there are either no imperiled vertebrae species or no cost estimates.

Source: USDA, Economic Research Service using data from NatureServe (2014) and the cost of new wetlands in figure 1.

Figure 8b. Number of wetland-associated imperiled* vertebrae species** relative to the cost of a new wetland, by county.

We estimated nitrogen-removal cost effectiveness in parts of the PPR. In these areas, 2 to 8 pounds of nitrogen are removed per dollar— $0.12 to $0.50 per pound—which is in the lower end of the $0.03 to $7.00 range estimated for wetlands across the Upper Mississippi and Ohio River Valleys. But, being different units of measure (e.g., pounds of nitrogen versus dollar benefits), the values are not additive.

We found no biophysical or economic data or models that would enable us to estimate wetlands' species-protection benefits. Spatial data indicate that the 57 percent of counties in the contiguous 48 States have at least one wetland-associated imperiled species. But restoring a wetland in any county, even in one that currently has no wetland-associated imperiled species, may affect species' survival probabilities.[28] Conversely, it is also possible that restoring a wetland in counties with many imperiled species may provide no support. Biophysical models that can predict wetlands' impacts on species' survival probabilities would be very insightful. Additionally, studies that estimate the relative strengths of the public's preferences (or their WTP) to protect different species are limited. As a result, it is difficult, at best, to evaluate the tradeoffs of restoring alternative wetlands.

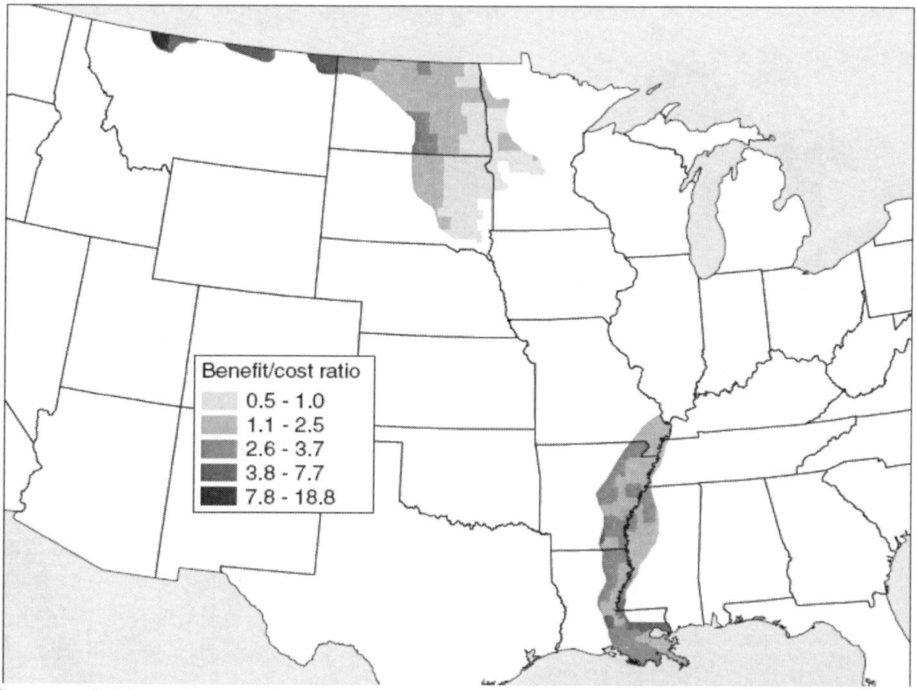

Source: USDA, Economic Research Service analysis of wetlands' impacts on greenhouse gases, (reported by Euliss et al. (2006) and the USFWS (2005)), estimated values of marginal changes in greenhouse gasses (reported by U.S. Government (2010)) and the wetland cost estimates in figure 1.

Figure 9. Wetlands' carbon sequestration benefits relative to wetland cost in the Prairie Pothole Region and the Mississippi Alluvial Valley, by county.

The Current Distribution of Restored and Protected Wetland Acreage

Although we did not estimate many wetland benefits, a comparison of benefits with the spatial distribution of wetland conservation acreages illustrates how the targeting of wetland conservation efforts might be improved. WRP contract acreage tends to be more concentrated in the MAV, the eastern portion of the PPR, and parts of California and Oregon (figure 11). We find CRP wetland- related contract acreage spread across much of rural America, but with considerable concentration in the PPR (figure 12). In a comparison with benefits, the most compelling findings include the following:

- Figure 11 indicates that, in the PPR, WRP acreage is denser in the eastern portion, where costs tend to be higher (see figure 2) and duck hunting benefits lower (see figure 3). Additionally, the total (hunting plus carbon sequestration) benefit-cost ratio estimates are higher where WRP contract density is lower (see figure 10). This suggests that, within the PPR, targeting wetland restoration and protection funding toward the western areas is likely to provide the greatest net social benefit— note that not all benefits have been estimated. Furthermore, while the imperiled species per-dollar estimates are high in the western portion of the PPR (see figure 8a), the value of wetlands' species protection is not necessarily higher.
- Figure 12 indicates that in the PPR, the CRP wetland-related contract acreage is denser in the eastern portion, where (as pointed out above) costs tend to be higher (see figure 2) and the duck hunting benefits lower (see figure 3). Because the CRP contracts are only 10 to 15 years, carbon sequestration benefits will be substantially less than the values reported here (unless the CRP-funded wetlands remain in place long after contracts expire) (see box "The Value of GHG Impacts Depends on GHG Emissions Over Time" on page 25).
- The MAV has some of the highest density of WRP acreage (see figure 11). The carbon sequestration benefits of the WRP acreage in most of this area tend to exceed costs—and these wetlands provide other benefits, including biodiversity and water quality. This result provides some justification for the greater density of the WRP acreage.

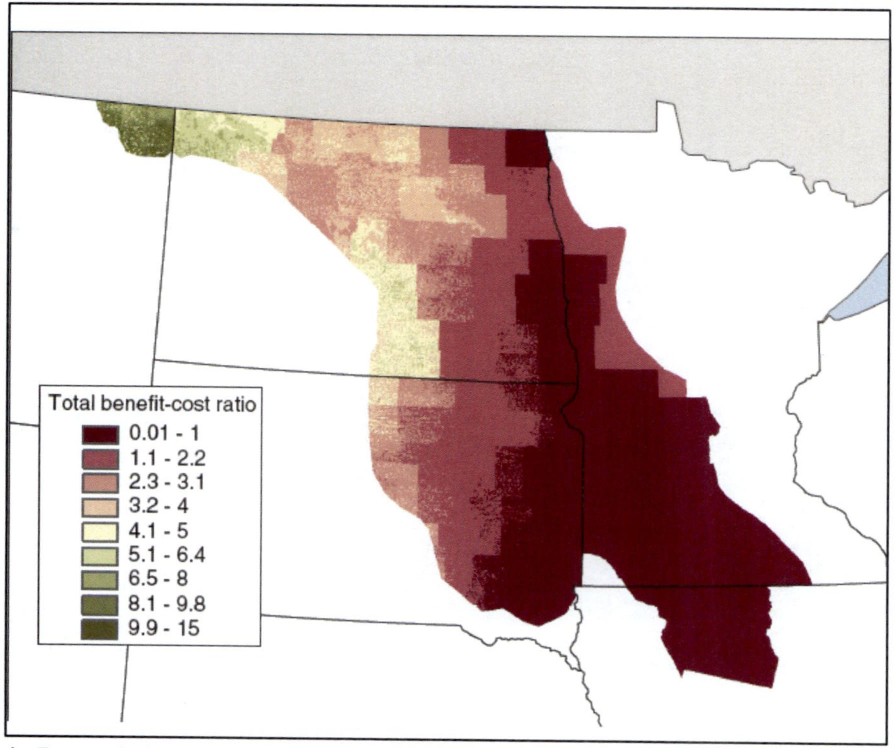

Source: USDA, Economic Research Service analysis of the benefits of duck hunting in figure 3, wetlands' impacts on greenhouse gasses, (reported by Euliss et al. (2006) and the USFWS (2005)), estimated values of marginal changes in greenhouse gasses (reported by U.S. Government (2010)), and the cost of new wetlands in figure 1.

Figure 10. Sum of duck-hunting and carbon-sequestration benefit-cost ratios of new wetlands in the Prairie Pothole Region.

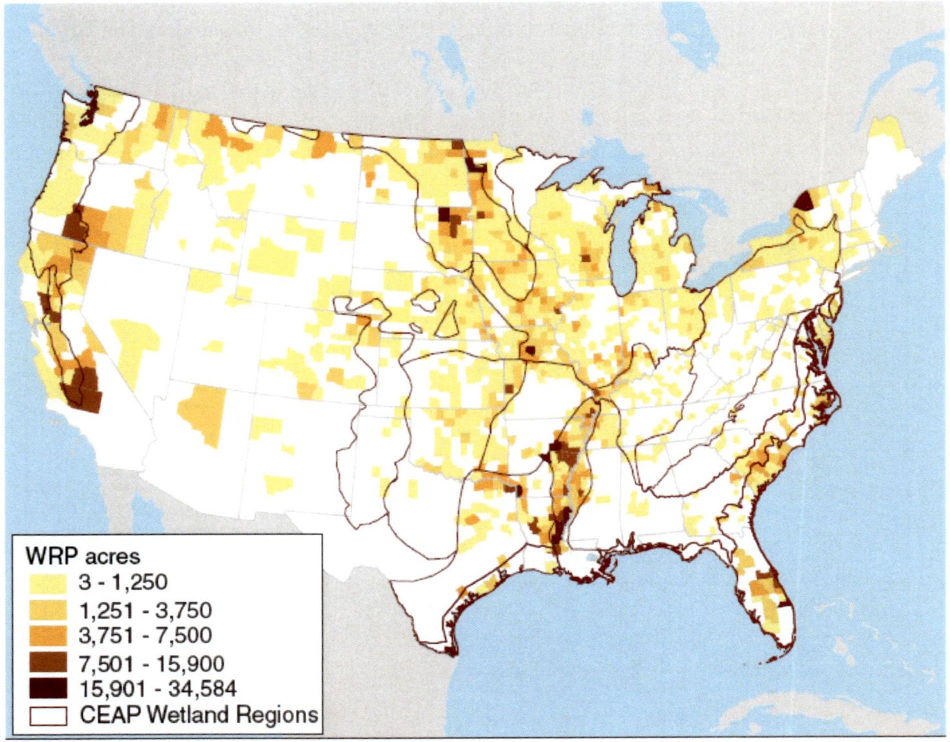

Source: USDA, Economic Research Service analysis of USDA, Natural Resources Conservation Service confidential data on Wetlands Reserve Program contracts.

Figure 11. Number of acres enrolled in the Wetlands Reserve Program, by county.

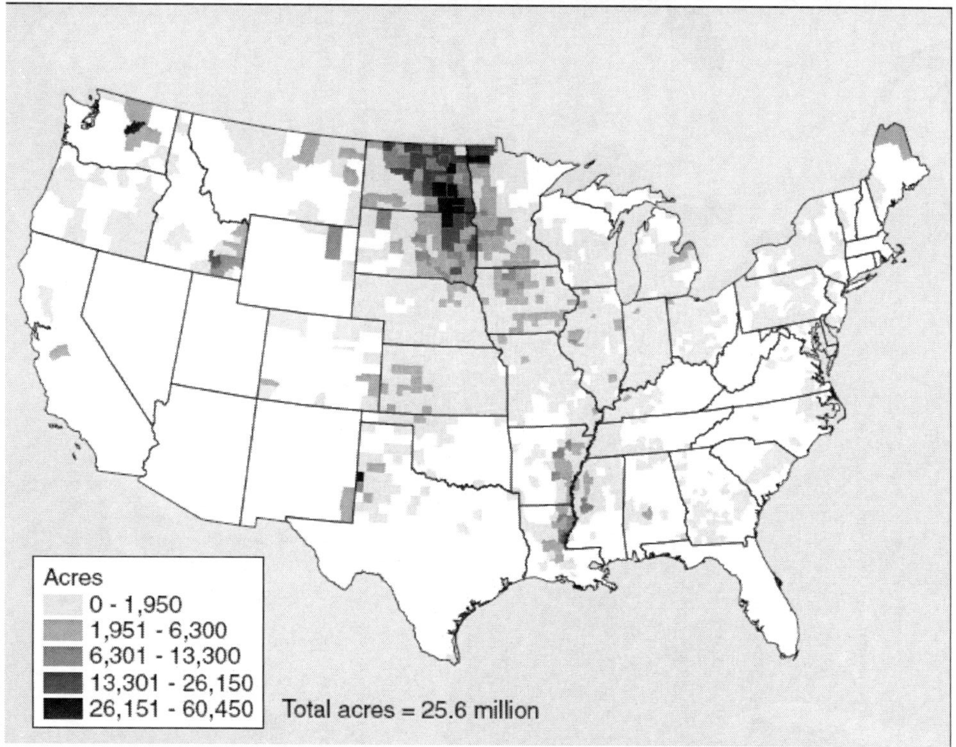

Source: USDA, Economic Research Service using confidential program data from USDA, Farm Service Agency, Conservation Reserve Program.

Figure 12. Distribution of Conservation Reserve Program wetland-related contract acreage (both wetland and upland), by county.

CONCLUSION

This study generates spatial estimates of the cost of restoring and preserving wetlands and the social values of some of the benefits they and their ecosystems produce. The results illustrate large spatial variations in wetland costs and benefits and economic advantages to targeting wetland conservation efforts. This information is designed to aid public policymakers, program operators, nongovern mental organizations, and the general public in efforts to employ evidence-based decisionmaking in wetland conservation. While limited by information constraints, this study begins to answer questions such as the following: Where are restored wetlands likely to be most cost effective at removing nitrogen? How much would duck hunters be willing to pay for additional wetlands across the PPR? How valuable are wetlands' greenhouse impacts? What is the likely cost of establishing a new wetland? Answers to these and similar questions are fundamental in constructing and improving conservation contract ranking criteria, such as environmental benefit indexes.

The findings and modeling framework discussed here provide guidance in setting future research priorities in two ways. First, and most apparent, the findings indicate where information is missing and draws attention to where further biophysical and economic research might be most useful. For example, we found that wetlands of the Upper Mississippi and Ohio River watersheds can be, but are not always, very effective at removing nitrogen, but we did not estimate the effectiveness of wetlands elsewhere. This result suggests that analyses of the nitrogen removal rates of wetlands in areas where nitrogen loadings impose social costs could prove to be useful.

Second, the study's analytic framework can help lay out relevant questions. First, is the value of the wetland benefit, $D(g(f(x)))$, likely to be large relative to other benefits? Second, how weak is the weakest link or what link(s) is(are) missing: The data on wetland characteristics: x? The models of wetland service: $f(x)$? The models of amenities produced: $g(f(x)))$? Or the demand or willingness to pay function: $D(g(f(x)))$? And third, how can the derivation of one model help narrow the focus of research on the other models? For example, before allocating research funding to estimate wetland effects on groundwater recharge, consider the economic importance of groundwater: Do many people in the area consume groundwater either directly or through irrigation applications? Will additional groundwater reduce the potential for seasonal shortages of stream flows? Also, consider other critical biophysical links: Are the affected aquifers the same as those that feed consumers and streams? If the answers to these questions are "no," then research efforts might be better channeled elsewhere.

After considering the contribution that economics can make and identifying some key information gaps, we offer suggestions for future research. This list is not comprehensive; rather, it is meant to illustrate areas where additional estimates will be useful.

- Duck hunting benefits were found to be high but the estimates drew on willingness-to-pay analyses that are dated. Because of the economic significance of duck hunting benefits, accuracy is important. An update of individuals' willingness to pay for improved hunting quality can help solidify results but will require data collection and analysis.
- Wetlands' duck hunting benefits were found to be high and variable, but the spatial extent of the biophysical models are limited. This suggests that research that expands the spatial extent of biophysical models might be useful.
- Dollar-per-ton estimates are available for some soil conservation benefits (Hansen and Ribaudo, 2008), and these estimates can be used to value wetlands' soil conservation benefits.[29] But, to do so, estimates of wetlands' effects on erosion are needed. One can expect erosion rates on converted wetland acreage to be less than county average rates because wetlands are commonly restored on flat lands. Land-use data, such as the National Resources Inventory, might be used to estimate erosion rates.
- The results of this study suggest that, in some areas, wetlands can be cost effective at removing nitrogen from nearby waterways. Thus, additional analyses of nitrogen removal by other sizes and types of wetlands across different-sized watersheds within and outside of the study area may be useful.
- We did not review literature on public consumption of groundwater—we did not have estimates of wetlands' effects on groundwater. Economic research that provides spatial estimates of the public's willingness to pay for increased groundwater will indicate to hydrologists where they might focus analyses of wetlands' effects on groundwater. Similarly, if research by hydrologists identifies areas where wetlands do affect groundwater, economists will see where to focus research on the valuation of groundwater.

- Similarly, economic research that generates spatial estimates of the social cost of algae blooms in freshwaters (due to phosphorus loadings) can suggest to biophysical scientists areas of the country where they might focus research on phosphorus removal by wetlands.
- The public's willingness to pay to protect species appears to be high but not infinite, as indicated by historical expenditures (Shogren et al., 1998). A further review of the economics literature along with a survey of peoples' preferences across species and levels of protection (e.g., changes in survival probabilities) can provide biologists with insights on those species that are of greatest public interest, hence, areas where they might prioritize research. The same information can also give those engaged in conservation efforts a better idea of how to prioritize funding for species protection. There are data that might enable analysts to examine the biophysical effects of wetlands on imperiled species. For example, information on the effects of habitat on species population may be contained within various descriptive text fields in NatureServe's central data-base (www.fort.usgs.gov/Products/ProdSpecies_Index.asp) and on the U.S. Geological Survey Web site (www.fort.usgs.gov/Products/ProdSpecies_Index.asp).

APPENDIX 1. THE COSTS OF WETLAND CONSERVATION OPTIONS

Our primary data are from a Wetlands Reserve Program (WRP) contract file. These data provide our associated variable: the total cost of a wetland (*TWC*). TWC is the sum of the restoration and easement costs of each contract. WRP contract data are agreed-upon contract payments and acreages and estimates of the wetland restoration costs—actual costs may differ. Overall, we expect these data to be representative of WRP costs. As long as the contract data are not systematically biased, they should be suitable for estimating wetland cost models. The data include a variable that identifies the county where the contract is located. We use this variable to link the WRP data to other data. We specified our cost model as:

$$TWC = f(Acres, AcresSq, AgrValue, AgrValueSq, Rural)$$

Information on the size of each contract (*Acres*) is included in the WRP contract file. As discussed earlier, we expect the cost of restoring wetlands to increase with the size of the wetland, all else being equal. And we expect this relationship to be nonlinear; hence, we have included the second-degree term acres squared (*AcresSq*). We also expect the nuisance costs of wetlands to be dependent on the size of the wetland and that the effect is nonlinear, all else being equal. We estimate each cost model using observations from the same wetland region to minimize variations in other determinants of restoration costs.

We have no county-level data on farmland values. But per-acre farmland rental rate data are available and can serve as a suitable proxy for variation in the value of the land's agricultural productivity. We generated indicators of the agricultural value of land in contracts (*AgrValue*), by multiplying the contract acreage (*Acres*) by the county-average dry land rental rates reported by USDA's National Agricultural Statistics Service (NASS). NASS generates rental values from various sources, including the recently initiated county-level cash rents survey (USDA, NASS, 2009). Because we expect a nonlinear relationship, we include the square of *AgrValue* (*AgrValueSq*).

We use the Population-Interaction Zones for Agriculture (PIZA) data to determine locations that may have urban pressure (*Rural*). The PIZA scores are designed to measure unban influence. The PIZA data classify counties as rural or as one of three levels of urban. (Documentation and PIZA data are available at: www.ers.usda.gov/data-products/population-interaction-zones-for-agriculture- (piza).aspx). We use the PIZA scores to generate a zero-one dummy variable, Rural, that equals one when urban influence is likely to be minimal.

Data that quantify actions taken to restore wetland hydrology and ecosystems and eliminate or control invasive species are not available. To overcome or minimize the effect of this shortcoming, we estimated separate cost functions for each of the USDA, Natural Resources Conservation Service's Conservation Effects Assessment Project (CEAP) wetland regions (see figure 1). By design, CEAP regions are defined by wetland similarities. It seems reasonable to assume that, within a wetland region, determinants of the cost of restoring a wetland—other than the size of the wetland—are less likely to vary than are determinants across wetlands regions.

We acknowledge that *AgrValue* and Acres are correlated, as the size of a contract is one of the drivers of the cost of land and the cost of restoring and improving a wetland. This correlation can weaken conclusions one can draw about the signs, sizes, and statistical significance (due to inflated standard errors) of the estimated coefficients. However, correlation between independent variables is a model structural issue: it affects one's ability to assess causality. It does not introduce systematic bias and therefore may not weaken the model's prediction capabilities, which is our motivation for estimating the cost functions.

Results and Model Applications

The adjusted R-squares of the estimated models top out at 0.88. Half capture more than 75 percent of the variation in the dependent variable; the three weakest capture 53 to 60 percent (table A1). Given the complications resulting from variable correlations, one cannot make any conclusions about the economic implications of the estimated coefficients.

Before generating estimates of expected TWC for a new wetland, we need to identify appropri ately representative contract sizes (Acres) for each wetland region. A comparison of the means and median indicates that the distribution of contract sizes is skewed toward the larger sizes (table A1). We chose to set Acres equal to the median-sized contract (for the region) as a useful representative measure since half of the contracts will be greater in size and half will be smaller. Because the cost functions are nonlinear, one does not know how cost estimates based on mean values compare with the potential range of estimated costs. For example, using the average wetland size is not likely to generate an estimate of average cost. One advantage of using median values is that the models will generate median cost estimates when based on median wetland sizes.

We generated estimates of total and average wetland costs for each county. Estimates show that there is less variability in projected costs in much of the Southeast than elsewhere; hence, benefit-cost ratios are less likely to be driven by variation in costs in the Southeast (see figure 2). The Appalachian Highlands— with a cost model that captures 88 percent of the variation in the dependent variable—has the widest range in predicted wetland costs, with values in a scattered distribution across the area. Predicted costs also vary widely across the Prairie Pothole Region, but the values show a consistent pattern of increase as one goes from the region's northwest counties to the southeastern counties.

Table A1. Estimated coefficients of the wetland cost models of the 10 wetland regions

Wetland region	Adjusted R-square	Variable*	Coefficient	Standard error**	Number of observations
Glaciated Interior Plains	0.81	Intercept	-1247.23	3162.32167	3,656
		AgrValue	13.323	0.58039	
		AgrValueSq	-2.98E-07	9.85E-07	
		Acres	434.2865	38.52275	
		AcresSq	-0.02602	0.00738	
Pacific Mountain	0.53	Intercept	-31659	27684	265
		AgrValue	46.83228	4.69902	
		AgrValueSq	0.000506	-0.00004666	
		Acres	-419.998	175.86181	
		AcresSq	0.37051	0.0653	
Prairie Pothole Region	0.79	Intercept	-3870.37	2615.75999	2,468
		AgrValue	29.22938	0.57392	
		AgrValueSq	-0.00016	0.00000509	
		Acres	-421.041	26.21381	
		AcresSq	0.19056	0.00684	
Gulf Atlantic Coastal Flats	0.58	Intercept	207399	71840	283
		AgrValue	-3.57109	1.7062	
		AgrValueSq	5.86E-06	9.89E-07	
		Acres	790.5241	102.37909	
		AcresSq	-0.02619	0.00321	
Appalachian Highlands	0.88	Intercept	6509.232	1167.2668	1,537
		AgrValue	40.12727	1.6565	
		AgrValueSq	-0.00016	0.00003148	
		Acres	-422.888	60.72724	
		AcresSq	0.23552	0.07277	
High Plains	0.71	Intercept	14978	8847.30368	240
		AgrValue	12.08406	3.76541	
		AgrValueSq	0.000264	0.00009641	
		Acres	473.1165	97.13946	
		AcresSq	-0.31101	0.05205	
Mississippi Alluvial Valley	0.60	Intercept	4339.446	9551.82475	1,417
		AgrValue	23.84496	2.49292	
		AgrValueSq	-5.7E-05	0.00001686	
		Acres	-417.548	108.46545	
		AcresSq	0.14741	0.02848	
Central Valley	0.76	Intercept	-69276	44584	199
		AgrValue	15.05214	9.40191	
		AgrValueSq	9.53E-05	0.00012050	
		Acres	1597.231	214.96889	
		AcresSq	-0.26999	0.07019	

Table A1. (Continued)

Wetland region	Adjusted R-square	Variable*	Coefficient	Standard error**	Number of observations
Gulf Atlantic Coastal Plains	0.70	Intercept	11896	11909	963
		AgrValue	8.59126	1.62986	
		AgrValueSq	2.86E-05	0.00000621	
		Acres	429.924	47.14001	
		AcresSq	-0.03638	0.00344	
Central Plains	0.84	Intercept	10650	4277.74641	706
		AgrValue	-3.57894	1.2879	
		AgrValueSq	0.000118	0.00000549	
		Acres	712.5908	43.61911	
		AcresSq	-0.11722	0.0055	

* In the first regression runs, we found that, in most cases, the coefficient on Rural was insignificant. Hence we dropped the variable. The R-square fell by less than two points.

** Despite the fact that our variables are likely to be correlated and, consequently, the standard errors of the coefficients inflated, nearly all variable coefficients are statistically significant at the 99-percent level.

Source: USDA, Economic Research Service.

APPENDIX 2. CONCEPTUAL FRAMEWORK FOR ASSESSING ENVIRONMENTAL BENEFITS

Economic and Biophysical Frameworks

The productivity of wetlands varies similar to the productivity of agricultural lands. There are different types of wetlands and different types of agricultural lands. Their outputs are affected by a variety of environmental factors. On the agricultural side, markets bring society's preferences and production resources into alignment: producers choose what, how, and how much to produce based on prices and factor productivity. The prices consumers accept reveal society's preferences.[30]

On the wetland side, there are few markets that can provide measures of the values of wetland services. Instead, efforts to cost effectively restore and protect wetlands must rely on estimates of wetland productivity and the public's demands.

This appendix provides a structural framework suitable to evaluating and comparing wetland conservation benefits. Consider the following simple algebraic formulation using *x, f(x), g(f(x)), and D (g(f(x)))*.

- *x:* measures of eco-environmental characteristics, such as the size and type of wetland, daily temperature and precipitation, climatic conditions, surrounding land uses, quantity and quality of inflow waters, and the existence of invasive species.
- f_i *(x)*: the quantity of service *i* that a wetland produces. This will be a function of eco-environmental characteristics. For example, changes in coastal wetland conditions affect offshore fishery populations.
  - $g_j(fi(x))$: the quantity of amenity *j* that is provided by a wetland. That is, a single service, f_i (*x*), can produce a variety of amenities. For example, a new wetland can yield an increase in fish populations, f_{FISH} (*x*), and the increase in fish populations can improve the catch of commercial fishing operations ($g_{HARVEST}$ (f_{FISH} (*x*))) and the quality of recreational fishing ($g_{FISHINGQUALITY}$(f_{FISH} (*x*))).[31]
  - $D_k(g_j(f_{i(x)}))$: an amenity can be of value to people and commercial operations. When amenities directly affect commercial markets, the measurement of value is straightforward—the value of a change in the amenity is equal to the resulting change in profits, $D_{PROFITS}$ ($g_{HARVEST}$(f_{FISH} *(x)*)). For example, extending the fish example, one could estimate the economic impact on commercial fishing by multiplying the dockside price of fish by the gain in harvest. When amenities affect nonmarket goods, estimating willingness to pay is more complicated. Again, extending the fish example, demand for improved fishing quality can be measured through a population survey where individuals are asked what they would be willing to pay for improved fishing quality. Or, in situations where participants are choosing from multiple sites with differences in fishing quality, willingness to pay can be derived from data on travel costs, fishing quality, and other environmental and population characteristics.

The boundaries between *x, f(x), g(f(x)), and D (g(f(x)))* are "grey." For example, one could define a wetland's effect on duckling or adult populations as a service. We specify the link between a wetland and its amenities in a two-step process, *g(f(x))* to emphasize that there can be a chain of relationships involved—that is, a new wetland removes nitrogen, which reduces nitrogen in river systems, which reduces nitrogen entering estuaries, which reduces hypoxia, which affects a bay's ecosystem, which affects marine fisheries.

The specifications of x, $f(x)$, $g(f(x))$, and $D\ (g(f(x)))$ are subject to the data and models that are available. Also, past environmental benefit analyses have applied shortcuts to measuring these relationships.

Alternative Approaches for Benefits Analyses

Reduced form models

Reduced form models commonly shortcut the need to estimate the effect of an environmental service on an amenity by modeling demand as a function of the service $D\ ((f(x))$. A variety of studies estimated regional and national water quality benefits of soil conservation practices by, first, using the Revised Universal Soil Loss Equation to estimate erosion, $f(x)$, and, second, estimating willingness to pay for changes in water quality as a function of soil erosion (Sullivan et al., 2004; Hansen et al., 2002; Feather et al., 1999). In a regional analysis of the Conservation Reserve Program's hunting benefits, Hansen et al. (1999) estimated a reduced-form spatial model where five of the independent variables (the surrounding mix of agricultural land uses) served as a proxy for the environmental amenity (hunting quality).

Proxy variables

Proxy or indicator variables can sometimes be substituted for variables or data that are not available. For example, one could use the observed percent change in fish catch rates in the fisheries industry as a proxy for the percent change in the quality of recreational fishing. The strength of a proxy variable hinges on the correlation between the proxy variable and the true variable's value.

Meta-analysis

Meta-analyses use findings and models reported in previous research as data. Meta-analyses have been used to directly value wetlands, sidestepping the need to estimate the many biophysical relationships and economic values of the amenities. A major shortcoming of meta analyses related to wetlands is that, in most cases, they provide a single national or a limited number of regional values, and, hence, capture virtually no spatially resolution to variations in wetland benefits. Another short-coming of meta analyses is that the studies used have not evaluated wetlands that are representative of the kinds of wetlands that are restored on agricultural land. For example, Brander et al. (2006) used the results of a study by Cooper and Loomis (1993) in a meta analysis of wetland values. But Cooper and Loomis used a travel cost analysis to determine hunters' willingness to pay to hunt sites on public lands. Additionally, willingness to pay is determined by the sizes and proximities of surrounding populations, water levels in the wetlands, and other factors—these factors and features of public wetlands are not likely to be representative of wetlands restored on agricultural lands.

Borisova-Kidder (2006) attempted to estimate a wetland-benefits model suitable for U.S. agriculture policy and program analyses but was unable to control for differences in wetland benefits across wetland types. For example, there was only one appropriate study that evaluated a wetland in the Prairie Pothole Region of the Upper Midwest.

APPENDIX 3. PREDICTING NEW WETLANDS EFFECTS ON DUCKS

We developed models that predict the number of hatchlings a new wetland is likely to produce for the five dominant duck species of the Prairie Pothole Region (PPRP (blue-winged teal (A*nas discors*), gadwall (*A. strepera*), mallard (*A. platyrhynchos*), northern pintail (*A. acuta*), and northern shoveler (*A. clypeata*)). Put simply, we model hatchlings (*H*) produced by a specific species at a wetland as a function of the nesting pairs (*NP*) the wetland attracts, nest success (*NS*), re-nesting propensity (*NI*), and clutch size (*CS*):

$$H = NP * NS * NI * CS.$$

We built this relationship from data on duck species, findings of prior research, and two types of models—a nesting pair model and a nest success model—that were developed for the PPR of North and South Dakota and northeastern Montana. We extended this model and drew on other resources to generate estimates of wetland effects on hatchlings in the PPR of Iowa and Minnesota.

We generated estimates of *NP* (nesting pairs) from a set of "pair-wetland relationship models," one for each species and wetland class (i.e., temporary, seasonal, semi-permanent, lake, river (Steward and Kantrud, 1971, 1972;

Cowardin et al., 1995) that were developed for field-level evaluations of the expected effects of a new wetland. Pair models were developed from breeding pair and wetland condition data collected during the U.S. Fish and Wildlife Service Annual Breeding Waterfowl Survey from 1987-2008 (Cowardin et al., 1995, Reynolds et al., 2006). We applied these pair models across 1 mile x 1 mile fishnet cells. This enabled us to generate a GIS surface of the number of breeding pairs added with a new wetland.

Nest success (*NS*) is the probability an egg survives all exposure days (Mayfield, 1975) and is a primary driver of duck population maintenance (Hoekman et al., 2002). Nest success is positively related to the share of surrounding grassland cover (Reynolds et al., 2001; Stephens et al., 2008). To estimate *NS*, we began with available daily nest survival rate (DNS) models (Mayfield, 1975) and developed spatially explicit models from Reynolds et al. (2001). The DNS model for a species$_i$ is expressed as $DNS_i = a_i + (b_1 * \textit{percent grass}) + (b^2 * \textit{easting}) + (b^3 * \textit{northing}) + (b_4 * \textit{easting} * \textit{northing})$, where a_i is the intercept, *percent grass* is the share of grass cover within a 1-mile radius of the centroid point of each pixel, and easting and northing are longitudinal and latitudinal measures of the centroid point (Reynolds et al., 2001).[32]

We used spatially specific measures of *percent grass from* landcover data from Thematic Mapper satellite imagery (93.5 ft. x 93.5 ft. pixel) for the years 2001-03 (Habitat and Population Evaluation Team, USFWS Region 6, Bismarck, ND, unpublished data). With these data and employing a moving window analysis, we calculated *percent grass* (which includes CRP) within the surrounding 3.14-mile^2 area (1 mile radius) of each pixel.

Finally, to generate spatial estimates of *NS* (the probability that a nest survives all days from when the first egg is laid to the day when the eggs hatch) for each species, we raised the *DNS* estimate of each species to the power of the number of nest-exposure days.

Nesting propensity (*NI*) accounts for cases where a nesting pair re-nests after losing its nest. By re-nesting, the nesting pair can still add to the hatchling population so it must be accounted for. Nesting propensity of a given species was treated as a constant and was taken for Bellrose (1980).

Average clutch size (*CS*) varies by species: mallard=8.4, gadwall=9.9, teal=10.2, pintail=9.9, and shoveler=7.1.

We generated estimates of the effect of new wetlands across the PPR and found that productivity ranges from near 0 to 57 hatchlings per acre. The distribution of the productivity of wetlands is skewed toward the lower end, with the bottom 10 percent of the wetlands likely to produce 3 hatch- lings per acre or less. The productivity of the top 10 percent is 18 hatchlings per acre or more. The median level of productivity is 5.5 hatchlings per acre.

APPENDIX 4. CALCULATING WETLANDS' NITROGEN REMOVAL POTENTIAL

We estimated the potential nitrate mass removal by wetlands as the product of nitrate mass load intercepted and the nitrate removal rate for locations across the Upper Mississippi and Ohio River Basins. Detailed descriptions on the estimation and application of the models that generated the mass loads and nitrate removal rates and supporting data sources are available in Crumpton et al. (2006) and Crumpton et al. (2010). In summary, Crumpton et al. (2006) generated spatial estimates of nitrate mass load based on the average water flow that would be intercepted by a wetland and the average nitrate concentration in that flow. Percent nitrate removal was calculated as a function of average hydraulic loading rate expected for wetlands at each location. Crumpton et al. (2010) extended the Crumpton et al. (2006) estimates to include the effect of tile drainage on nitrate concentration. Essentially, they estimated field nitrogen mass loss within a wetland watershed as a function of land uses (fraction of area in corn and soybean cultivation), tile drainage (fraction of crop land with tile drainage), and hydraulic loading rate and found all variables to be statistically significant (Crumpton et al., 2010).

Our nitrogen-removal estimates were generated using the above models, Landsat land cover data from Landsat Thematic Mapper imagery, NRCS Soil Survey Geographical (SSURGO) data, 1980-2005 U.S. Geological Survey stream gauge data, and 20-year simulations to capture expected variations in precipitation and rainfall intensity and their subsequent impact on nitrogen movement and wetland effectiveness. We assumed that wetlands are strategically placed at the bottoms of 1,000- acre wetland watersheds.[33] Our review of the literature and the analyses supporting the models we have applied indicate that a 1,000-acre wetland watershed is likely to be large enough to supply enough water to maintain a healthy wetland ecosystem and not atypical within our study areas. Furthermore, a recent study found that, in northwestern Ohio, a wetland of roughly 10 acres is most cost effective at nitrogen removal (Hansen et al., 2012). While other sized wetlands and wetland watersheds might have been considered, for the sake of brevity, we have not done so here. Our results indicate that, depending on where it is located, a new 10-acre wetland may remove 11 pounds to nearly 1 ton of nitrogen per wetland acre (see figure 4).

REFERENCES

Ando, A., Camm, J., Polasky, S. & Solow. A. (1998). "Species Distributions, Land Values, and Efficient Conservation," *Science*, *279*(5359), 2126-2128.

Arnold, J. & Fohrer. N. (2005). "SWAT 2000: Current Capabilities and Research Opportunities in Applied Watershed Modeling," *Hydrological Processes*, *19*(3), 563-72.

Batt, B., Anderson, M., Anderson, M. & Caswell. F. (1989). "The Use of Prairie Potholes by North American Ducks," 204-27, in A.G. van der Valk (ed.), *Northern Prairie Wetlands*, Ames, IA: Iowa State University Press.

Bengtson, M. & Padmanabhan. G. (1999). *A Review of Models for Investigating the Influence of Wetlands on Flooding*, North Dakota Water Resources Research Institute, North Dakota State University.

Bernal, B. & Mitsch. W. (2012). "Comparing Carbon Sequestration in Temperate Freshwater Wetland Communities," *Global Change Biology*, *18*(5), 1636-47.

Bianchi, T., DiMarco, S., Cowan Jr., J., Hetland, R., Chapman, P., Day, J. & Allison. M. (2010). "The Science of Hypoxia in the Northern Gulf of Mexico: A Review," *Science of the Total Environment*, *408 (7)*, 1471-84.www.sciencedirect.com /science/article/pii/S0048969709011620 (September, 2014).

Borisova-Kidder, A. (2006). "Meta-Analytical Estimates of Values of Environmental Services Enhanced by Government Agricultural Conservation Programs," Ph.D. Dissertation, Ohio State University.

Brander, L. M., Florax, R. & Vermaat. J. (2006). "The Empirics of Wetland Valuation: A Comprehensive Summary and a Met-Analysis of the Literature," *Environmental and Resource Economics*, *33*, 223-50.

Bridgham, S., Megonigal, J., Keller, J., Bliss, N. & Trettin. C. (2006). "The Carbon Cycle of North American Wetlands," *Wetlands*, *26*(4), 889-916.

Brown, G. Jr., Charbonneau, J. & Hay. M. (1978). *"Estimating Values of Wildlife: An Analysis of the 1975 Hunting and Fishing Survey,"* Working Paper No. 7, U.S. Department of the Interior, Fish and Wildlife Service, Division of Program Plans.

Brown, G. Jr., & Hay. M. (1987). *"Net Economic Recreation Values for Deer and Waterfowl Hunting and Trout Fishing, 1980,"* Working Paper No. 23. U.S. Department of the Interior, U.S. Fish and Wildlife Service, Division of Policy and Directives Management.

Bullock, A. & Acreman. M. (2003). "The Role of Wetlands in the Hydrological Cycle," *Hydrology and Earth System Science*, *7*, 358-89.

Charbonneau, J. & Hay. M. (1978). "Determinants and Economic Values of Hunting and Fishing," Transactions of the 43rd North American Wildlife and Natural Resources Conference, 391-403 Wildlife Management Institute, Washington, DC.

Costanza, R., Perez-Maqueo, O., Luisa Marinez, M., Sutton, P., Anderson, S. J. & Mulder. K. (2008). "The Value of Coastal Wetlands for Hurricane Protection," *AMBIO: A Journal of the Human Environment*, *37(4)*,*241-8.*

Cooper, J. & Loomis. J. (1993). "Testing Whether Waterfowl Hunting Benefits Increase With Greater Water Deliveries to Wetlands," *Environmental and Resource Economics*, *3*, 545-61.

Cowardin, L. M.,. Shaffer, T. L. & Arnold. M. (1995). *Evaluations of Duck Habitat and Estimation of Duck Population Sizes With a Remote-Sensing-Based System*, Biological Science Report 2, National Biological Service, Northern Prairie Wildlife Research Center, Jamestown, ND.

Crumpton, W., Green, D., Loan-Wilsey, A., McNeely, R., Kane, K., Otis, D. T. Cooper, D. & Johnson. R. (2010). *Assessment of Environmental Services of CREP Wetlands in Iowa and the Midwestern Corn Belt*, Progress Report IV, Prepared for U.S. Department of Agriculture, Farm Service Agency.

Crumpton, W., Stenback, G., Miller, B. & Helmers. M. (2006). *Potential Benefits of Wetland Filters for Tile Drainage Systems: Impact on Nitrate Loads to Mississippi River Subbasins*, CSREES Project Completion Report, U.S. Department of Agriculture, Cooperative State Research, Education, and Extension Service.

Denver J., Ator, S., Lang, M., Fisher, T., Gustafson, A., Fox, R., Clune, J., & McCarty. G. (2014). "Nitrate Fate and Transport Through Current and Former Depressional Wetlands in an Agricultural Landscape, Choptank Watershed, Maryland, United States," *Journal of Soil and Water Conservation*, *69*(1), 1-16.

Diaz, R. & Solow. A. (1999). *Ecological and Economic Consequences of Hypoxia: Topic 2 Report for the Integrated Assessment on Hypoxia in the Gulf of Mexico*, NOAA Coastal Ocean Program Decision Analysis Series No. 16, U.S. Department of Commerce, National Oceanic and Atmospheric Administration, Coastal Ocean Program.

Dobson, A. P., Rodriguez, J. P., Roberts, W. M. &. Wilcove. D. S. (1997). "Geographic Distribution of Endangered Species in the United States," *Science*, *257*, 550-3.

Doss, C., & Taff. S. (1996). "The Influence of Wetland Type and Wetland Proximity on Residential Property Values," *Journal of Agricultural and Resource Economics*, *21*(1), 120-9.

Eagle, A., Olander, L., Henry, L. R., Haugen-Kozyra, K., Millar, N. & Robertson. G. P. (2012). *Greenhouse Gas Mitigation Potential of Agricultural Land Management in the United States: A Synthesis of the Literature*, Report NIR 10-04, Third Edition, Nicholas Institute for Environmental Policy Solutions, Duke University, Durham, NC.

Earnhart, D. (2001). "Combining Revealed and Stated Preference Methods to Value Environmental Amenities at Residential Locations," *Land Economics*, *77*(1), 12-29.

Euliss, N. Jr., Gleason, R., Olness, A., McDougal, R., Murkin, H., Robarts, R., Bourbonniere, R. & Warner. B. (2006). "North American Prairie Wetlands Are Important Non-Forested Land-Based Carbon Storage Sites," *Science of the Total Environment*, *361*, 179-88.

Farber, Stephen. (1987). "The Value of Coastal Wetlands for Protection of Property Against Hurricane Wind Damage," *Journal of Environmental Economics and Management*, *14*(2), 143-151.

Faulkner, S. Barrow, W., Keeland, B., Walls, S., Moorman, T., Twedt, D. & Uihlein, W. III. (2008). *Assessment of Ecological Services Derived from U.S. Department of Agriculture Conservation Programs in the Mississippi Alluvial Valley: Regional Estimates and Functional Condition Indicator Models*, Interim Report, U.S. Geological Survey, National Wetlands Research Center.

Feather, P., Hellerstein, D. & Hansen. L. (1999). *Economic Valuation of Environmental Benefits and the Targeting of Conservation Programs: The Case of the CRP*, Agriculture Economic Report No. 778, U.S. Department of Agriculture, Economic Research Service, www.ers.usda.gov/publications/aer-agricultural-economic-report/aer778.aspx#.U_4gqWM_xQI

Freeman, A., III. (1993). *The Measurement of Environmental and Resource Values: Theory and Methods*, Washington, DC: Resources for the Future.

Gascoigne, W., Hoag, D., Koontz, L., Tangen, B., Shaffer, T. & Gleason. R. (2011). "Valuing Ecosystem and Economic Services Across Land-Use Scenarios in the Prairie Pothole Region of the Dakotas, USA," *Ecological Economics*, *70*(10), 1715-25.

Gleason, R. & Euliss Jr. N. (1998). "Sedimentation of Prairie Wetlands," *Great Plains Research*, *8*(1), Northern Prairie Wildlife Research Center Online, www.npwrc.usgs.gov/resource/wetlands/sediment/index.htm

Gleason, R., Euliss, N. Jr., Mcdougal, R., Kermes, K., Steadman, E. & Harju. J. (2005). *Potential of Restored Prairie Wetlands in the Glaciated North American Prairie to Sequester Atmospheric Carbon*, Plains CO_2 Reduction Partnership, University of North Dakota.

Gleason, R., Laubhan, M. & Euliss, N. Jr. (eds). (2008). *Ecosystem Services Derived from Wetland Conservation Practices in the United States Prairie Pothole Region with an Emphasis on the U.S. Department of Agriculture Conservation Reserve and Wetland Reserve Programs*, Professional Paper 1745, U.S. Department of the Interior, U.S. Geological Survey.

Gleason, R., Tangen, B., Browneb, B. & Euliss, N. Jr. (2009). "Greenhouse Gas Flux From Cropland and Restored Wetlands in the Prairie Pothole Region," *Soil Biology & Biochemistry*, *41*, 2501-7.

Gupta, T. & Foster. J. (1975). "Economic Criteria for Freshwater Wetland Policy in Massachusetts," *American Journal of Agricultural Economics*, *57*, 40-45.

Hammack, J. & Brown, G. M. Jr. (1974). *Waterfowl and Wetlands: Toward Bioeconomic Analysis*, Washington, DC. Resource for the Future, Inc.

Hansen, L. & Ribaudo. M. (2008). *Soil Conservation Benefit Variables: Regional Values for Policy Assessment*, Technical Bulletin No. 1922, U.S. Department of Agriculture, Economic Research Service, September, www.ers.usda.gov/publications/tb-technical-bulletin/tb1922.aspx#.U_4gAGM_xQI

Hansen, L., Delgado, J., Ribaudo, M. & Crumpton. W. (2012). "Minimizing Costs of Reducing Agricultural Nitrogen Loadings: Choosing Between On- and Off-field Conservation Practices," *Environmental Economics*, *3*(4), 97-112.

Hansen, L., Feather, P. & Shank. D. (1999). "Valuation of Agriculture's Multisite Environmental Impacts," *Agricultural and Resource Economics Review*, *28*(2), 199-207.

Hansen, L., Breneman, V., Davison, C. & Dickens. C. (2002). "The Cost of Soil Erosion to Downstream Navigation," *Journal of Soil and Water Conservation 57(4), 205-212.*

Hoekman S., Mills, L., Howerter, D., Devries, J. & Ball. I. (2002). "Sensitivity Analyses of the Life Cycle of Midcontinent Mallards," *Journal of Wildlife Management*, *66*, 883-900.

Irwin, E. & Bockstael. N. (2004). "Land Use Externalities, Open Space Preservation, and Urban Sprawl," *Regional Science and Urban Economics*, *34*, 705-725.

Jenkins, W. Aaron, Brian C. Murray., Randall A. Kramer., Stephen, E. & Faulkner, P. (2010). "Valuing Ecosystem Services From Wetlands Restoration in the Mississippi Alluvial Valley," *Ecological Economics*, *69*(5), 1051-1061.

Kaul, I., Grunberg, I. & Stern. M. (1999). "Introduction," in *Global Public Goods: International Cooperation in the 21st Century*, Inge Kaul, Isabellle Grunberg, and Marc A. Stern (eds)., New York: Oxford University Press.

Kayranli, B., Scholz, M., Mustafa, A. & Hedmark. Å. (2010). "Carbon Storage and Fluxes Within Freshwater Wetlands: A Critical Review," *Wetlands* 30, 111-124.

Kazak, N. & BenDor. T. (2013). "The Land Value Impacts Wetland Restoration," *Journal of Environmental Management*, Vol. *127*, 289-299, September.

Krapu, G., Pietz, P., Brandt, D. & Cox, R. Jr. (2000). "Mallard Brood Survival in Prairie Pothole Landscapes," *Waterfowl*, *13*, 18.

Langpapa, C. & Kerkvliet. J. (2012). "Endangered Species Conservation on Private Land: Assessing the Effectiveness of Habitat Conservation Plan," *Journal of Environmental Economics and Management*, *64*(1), 1-15.

Larson, D. & Lew. D. (2014). "The Opportunity Cost of Travel Time as a Noisy Wage Fraction," *American Journal of Agricultural Economics*, *96*(2), 420-437.

Lawley, C. & Towe. C. (2014). "Capitalized Costs of Habitat Conservation Easements," *American Journal of Agricultural Economics*, *96*(3), 657-672.

Lichtenberg, E. (2011). "Open Space and Urban Sprawl: The Effects of Zoning and Forest Conservation Regulations in Maryland," *Agricultural and Resource Economics Review*, *40*(3), 393-404, December.

Loomis, J. & Walsh. R. (1997). *Recreation Economic Decisions*, *Comparing Benefits and Costs*, State College, PA: Venture Publishing, Inc., 436 p.

Loomis, J. & White. D. (1996). "The Economics Benefits of Rare and Endangered Species: Summary and Meta-Analysis," *Ecological Economics*, *18*, 197-206.

Ma, S. & Swinton. S. (2011). "Valuation of Ecosystem Services from Rural Landscapes Using Agricultural Land Prices," *Ecological Economics*, *70*, 1649-1659.

MacKenzie, J. (1993). "A Comparison of Contingent Preference Models," *American Journal of Agricultural Economics*, *75*, 593-603, August.

Mahan B. Polasky, S. & Adams. R. (2000). "Valuing Urban Wetlands: A Property Price Approach," *Land Economics*, *76*(1), 100-113.

Mayfield, H. (1975). "Suggestions for Calculating Nest Success" *Wilson Bulletin*, *87*, 456-466. Menon, M., F. Perali, and M. Veronesi. 2014. "Recovering Individual Preferences for Non-Market Goods: A Collective Travel-Cost Model," *American Journal of Agricultural Economics*, *96*(2), 438-457.

Mississippi River Gulf of Mexico Watershed Nutrient Task Force. (2013). *Reassessment 2013: Assessing Progress Made Since 2008*, U.S. Environmental Protection Agency, Office of Wetlands, Oceans, and Watersheds.

Mitsch, W., Bernal, B., Nahlik, A., Mander, U., Zhang, L., Anderson, C. Jørgensen, S. & Brix. H. (2012). "Wetlands, Carbon, and Climate Change," *Landscape Ecology*, Published online: DOI 10.1007/ s10980-012-9758-8

NatureServe. (2014). "NatureServe's Central Databases," Arlington, VA.

Nichols, J. & Hines. J. (1983). "The Relationship Between Harvest and Survival Rates of Mallards: A Straightforward Approach With Partitioned Data Sets," *Journal of Wildlife Management*, *47*, 334-348.

Nordhaus, W. (2014). "Estimates of the Social Cost of Carbon: Concepts and Results From the DICE-2013R Model and Alternative Approaches," *Journal of the Association of Environmental and Resource Economists*, *1*(1/2).

Pattison, J. Boxall, P. & Adamowicz. W. (2011). "The Economic Benefits of Wetland Retention and Restoration in Manitoba," *Canadian Journal of Agricultural Economics/Revue Canadienne d'agroeconomie*, *59*, 223-244.

Pease, J.,. Rankin, M. L., Verdon, J. & Reisz. R. (1997). "Why Landowners Restore Wetlands: A National Survey," Iowa State University Extension, Department of Animal Ecology.

Rabalais, N. & Turner (eds). R. (2001). "Coastal Hypoxia: Consequences for Living Resources and Ecosystems," American Geophysical Union, Washington, DC.

Rabotyagov, S., Campbell, T., Philip M., Gassman, W., Arnold, J., Kurkalova, L., Secchi, S. & li, Feng H. & Kling. C. (2010). "Least-Cost Control of Agricultural Nutrient Contributions to the Gulf of Mexico Hypoxic Zone, *Ecological Application*, *20*(6), 1542-1555, http://dx.doi. org/10.1890/08-0680.1

Rabotyagov, S. Kling, C., Gassman, P., Rabalais, N. & Turner. R. (2014). "The Economics of Dead Zones: Causes, Impacts, Policy Challenges, and a Model of the Gulf of Mexico Hypoxic Zone," *Review of Environmental Economic Policy*, *8*(1), 58-79, doi: 10.1093/reep/ret024

Regan, T., Master, L. & Hammerson. G. (2004). "Capturing Expert Knowledge for Threatened Species Assessments: A Case Study Using NatureServe Conservation Status Ranks," *Acta Oecologica*, *26*, 95-107.

Reynolds, R. & Regalado. A. (2002). "The Effects of Wetlands and Other Factors on Rural Land Values," *Appraisal Journal*, *72*, 182-190, April.

Reynolds, R., Loesch, C., Wangler, B., Shaffer. T. (2007). "Waterfowl Response to the Conservation Reserve Program and Swampbuster Provision in the Prairie Pothole Region, 1992-2004," U.S. Department of the Interior, Bismarck, ND.

Reynolds R., Shaffer, T., Loesch, C. & Cox, R. Jr. (2006). "The Farm Bill and Duck Production in the PPR: Increasing the Benefits," *Wildlife Society Bulletin*, *34*(4), 963-974, November.

Reynolds R., Shaffer, T., Renner, R., Newton, W. & Batt. B. (2001). "Impact of the Conservation Reserve Program on Duck Recruitment in the U.S. Prairie Pothole Region," *Journal of Wildlife Management*, *65*, 765 780.

Richardson, L. & Loomis. J. (2009). "The Total Economic Value of Threatened, Endangered and Rare Species: An Updated Meta-Analysis," *Ecological Economics*, *68*, 1535-1548.

Ribaudo, M. (1989). *Water Quality Benefits from the Conservation Reserve Program*, U.S. Department of Agriculture, Economic Research Service, Agricultural Economic Report No. 606, February, www.ers.usda.gov/publications/aer-agricultural-economic-report/aer606.aspx#.U_4k7mM_xQI

Ribaudo, M., Delgado, J., Hansen, L., Livingston, M., Mosheim, R., Williamson. J. (2011). *Nitrogen in Agricultural Systems: Implications for Conservation Policy*, Economic Research Report No. 127, U.S. Department of Agriculture, Economic Research Service, www.ers.usda.gov/publications/err- economic-research-report/err127.aspx

Ribaudo, M., Hansen, L., Hellerstein, D. & Greene, C. (2008), *The Use of Markets To Increase Private Investment in Environmental Stewardship*, Economic Research Report No. 64, U.S. Department of Agriculture, Economic Research Service, September, www.ers.usda.gov/publications/err-economic-research-report/err64.aspx#.U_4lmGM_xQI

Rosen, S. (1974). "Hedonic Prices and Implicit Markets: Product Differentiation in Pure Competition," *Journal of Political Economy*, *82*(1), 34-55.

Roughgarden, J. (1995). *The Economics and Ecology of Biodiversity Decline*, Cambridge, UK: Cambridge University Press.

Shogren, J. F., Tschirhart, J., Anderson, T., Whritenour, A., Ando, A., Beissinger, S. R., Brookshire, D., Brown, G. M., Coursey, Jr., D., Innes, R. Meyer, S. M. & Polasky. S. (1998). "Why Economics Matters for Endangered Species Protection and the ESA," *Conservation Biology*, *3*(6), 1257-61.

Shultz, S. & Leitch. J. (2003). "The Feasibility of Restoring Previously Drained Wetlands to Reduce Flood Damage," *Journal of Soil and Water Conservation*, *58(1), 21-29.*

Smith, L., Haukos, M., McMurry, S., LaGrange, T. & Willis. D. (2011). "Ecosystem Services Provided by Playas in the High Plains: Potential Influences of USDA Conservation Programs," *Ecological Applications 21*, S82-S92.

Smith, K. & Weinberg. M. (2004). "Measuring the Success of Conservation Programs," *Amber Waves*, USDA, Economic Research Service, September, http://webarchives.cdlib.org/sw1vh5dg3r/ http:/ers.usda.gov/ amberwaves/september04/features/measuringsuccess.htm

Stephens, S., Walker, J., Blunck, D., Jayaraman, A. Nalugle, D., Ringelman, J. & Smith. A. (2008). "Predicting Risk of Habitat Conversion in Native Temperate Grasslands," *Conservation Biology*, *22*, *1320-1330*, *doi: 10.1111/j.1523-1739.2008.01022.x*

Stewart, R. E. & Kantrud. H. A. (1971). *Classification of Natural Ponds and Lakes in the Glaciated Prairie Region*, Resource Publication 92, U.S. Fish and Wildlife Service, www.npwrc.usgs.gov/ resource/wetlands/ pondlake/ index.htm

Sullivan, P., Hellerstein, D., Hansen, L., Johansson, R., Koenig, S., Lubowski, R., McBride, W., McGranahan, D., Roberts, M., Vogel, S. & Bucholtz. S. (2004). *The Conservation Reserve Program: Economic Implications for Rural America*, Agricultural Economic Report No. 834, U.S. Department of Agriculture, Economic Research Service, October, www.ers.usda.gov/ publica-tions/aer-agricultural-economic-report/aer834.aspx#.U_ 4oJ2M_xQI

Szentandrasi, S., Polasky, S., Berrens, R. & Leonard. J. (1995). "Conserving Biological Diversity and the Conservation Reserve Program," *Growth & Change*, *26*(3), 383-404.

U. S. Army Corps of Engineers. (2013). "Conservation Reserve Program Flood Damage Reduction Benefits to Downstream Urban Areas," Rock Island District Pilot Project Report, Mississippi Valley Division Initiative, www.fsa.usda.gov/Internet/FSA_File/fsa_pilot_report_final.pdf

U.S. Department of Agriculture, National Agricultural Statistics Service (USDA, NASS). (2009). "Land Values and Cash Rents: 2009 Summary," August, http://usda.mannlib.cornell.edu/usda/ nass/AgriLandVa//2000s/ 2009/AgriLandVa-08-04-2009.pdf.

U. S. Department of Commerce, National Oceanic and Atmospheric Administration. (2008). "Billion Dollar U.S. Weather Disasters 1980-2008," National Climatic Data Center, www.ncdc.noaa.gov/oa/ reports/billionz.html

U. S. Environmental Protection Agency. (2014). "Benefits Analysis for the Final Section 316(b) Existing Facilities Rule," EPA-821-R-14-005, May, http://water.epa.gov/lawsregs/lawsguidance/ cwa/316b/upload/Cooling-Water_Phase-4_Benefits_2014.pdf

U. S. Environmental Protection Agency, Science Advisory Board (SAB). (2007). "Hypoxia in the Northern Gulf of Mexico: An Update," EPA-SAB-08-003, http://yosemite.epa.gov/sab/ sabproduct.nsf/ C3D2F27094E03F 90852573B800601D93/$File/EPA-SAB-08-003complete. unsigned.pdf

U. S. Federal Reserve. (2012). "Moody's Seasoned AAA Corporate Bond Yield," Board of Governors of the Federal Reserve System, www.forecasts.org/index.htm.

U. S. Fish and Wildlife Service. (2012). "Adaptive Harvest Management: 2012 Hunting Season," U.S. Department of Interior, Washington, DC, www.fws.gov/migratorybirds/mgmt/AHM/ AHM-intro.htm

U.S. Fish and Wildlife Service. Habitat and Population Evaluation Team, USFWS Region 6, Bismarck, ND, unpublished data.

U.S. Fish and Wildlife Service. (2005). "Entergy, Louisiana Partners Celebrate 2,900-Acre Addition to Tensas River Refuge Using Innovative Conservation Tool," Press release May 23, www.fws. gov/southeast/news/ 2005/r05-041.html

U. S. Government. (2010). "Interagency Working Group on Social Cost of Carbon," Technical Support Document: Social Cost of Carbon for Regulatory Impact Analysis Under Executive Order 12866, www.epa.gov/otaq/climate/regulations/scc-tsd.pdf

Van Kooten, Cornelis, G. Withey, P. & Wong. L. (2011). "Bioeconomic Modeling of Wetlands and Waterfowl in Western Canada: Accounting for Amenity Values," *Canadian Journal of Agricultural Economics*, January.

Whitehead, J. (2005). “Environmental Risk and Averting Behavior: Predictive Validity of Jointly Estimated Revealed and Stated Behavior Data,” *Environmental and Resource Economics*, *32*, 301-316.

Woodward, R. & Wui. Y. (2001). “The Economic Value of Wetland Services: A Meta Analysis,” *Ecological Economics*, *37*, 257-270.

Zinn J. & Copeland. C. (2007). *Wetlands: An Overview of Issues*, Congressional Research Report, Congressional Research Service, Washington, DC.

End Notes

[1] The Agricultural Act of 2014 rolled functions of the Wetlands Reserve Program, Grassland Reserve Program (easement portion), and the Farm and Ranch Lands Protection Program into the Agricultural Conservation Easement Program (ACEP). All land enrolled in these earlier programs is now considered enrolled in ACEP. The three earlier programs received average annual funding of $686 million for 2008-13, but annual ACEP funding for 2014-18 is expected to average $375 million.

[2] Total WRP acreage data are from www.nrcs.usda.gov/Internet/NRCS_RCA/ reports/ fb08_cp_wrp.html. The breakdown of Wetlands Reserve Program (WRP) contract acreages is based on ERS analysis of detailed data on WRP contracts supplied by USDA’s Natural Resources Conservation Service. The detailed contract data file has a little more than half of all WRP contracts. We have assumed that the distribution of contract types in it is representative of all WRP contracts.

[3] Estimates are based on CRP data supplied by USDA’s Farm Service Agency and include only yearly rental payments (cost share expenses for installing land covers, and the cost of signup incentive payments, are not included).

[4] At the end of fiscal year 2014, the CRP had 2 million wetland-related acres under contract (http://www.fsa.usda.gov/ Internet/FSA_File/summarysept2014.pdf). At the time of publication of this report, WRP contract data were not available beyond 2012.

[5] This result is based on an analysis of all CRP contracts. Current data limitations do not allow for an evaluation of expiring wetland-related CRP contracts.

[6] The general CRP uses a competitive auction mechanism to select acres. However, most CRP wetland contracts are enrolled via the continuous CRP, which accepts any offer that meets eligibility requirements.

[7] Past research on wetland benefits has failed to recognize the value of the land with wetlands. See, for example, Jenkins et al. (2010) and van Kooten et al. (2011).

[8] Studies provide strong evidence that, in the past, demand for lands with wetlands can sometimes drive the value of the lands above their agricultural and alternative-use values (Reynolds and Regalado, 2002: Earnhart, 2001; Pease et al., 1997).

[9] We included a variable that was meant to account for the effects of urban proximity, found it to be statistically insignificant, and dropped it from the final model. Our belief is that urban proximity has offsetting effects. The potential for urban development can boost farmland prices above their agricultural values (which will tend to increase the cost of an easement). Conversely, urban proximity can increase the values of land and surrounding wetlands because of greater demand for wetland amenities (see appendix 1).

[10] Lawley and Towe (2014) recently examined the cost of easements on existing wetlands that cap land uses at the current levels in the Prairie Pothole Region (PPR) of Canada. They found that easement payments averaged $86 per acre (U.S. $). Two primary factors suggest

that, although low, the Lawley and Towe (2014) estimate is consistent with the ERS estimate in this study. First, the easements in this analysis are placed on land that had been drained and is being cropped, while Lawley and Towe (2014) evaluated easements on existing wetlands (e.g., land in a lower valued use). Second, the estimated pattern of easement costs (they drop as one moves northwesterly) in this study suggests that easement costs would be lower in Canada (in an area that is north and west of the U.S. PPR).

[11] For example, see the U.S. Environmental Protection Agency's discussion on wetland impacts at: http://water.epa.gov/type/wetlands/outreach/upload/EconomicBenefits.pdf

[12] Not all wetland "benefits" are positive—some people may look at wetlands as a source of mosquitos and displeasing odor.

[13] We apply a 4-percent market rate of discount in this analysis, the longrun return on AAA bonds (U.S. Federal Reserve, 2012), because interest rates observed on AAA bonds are reflective of the time preference for money, void of any risk premium.

[14] Valuing changes in amenities is complicated by the nonmarket nature of these goods. Instead of market prices, valuation models are used to calculate a marginal price. (See box "Analytical Methods for Valuing Environmental Amenities" for a short list and brief description of methods used to value environmental amenities.)

[15] Knowledge of these links is also necessary when evaluating benefits of on-field conservation incentives. Additionally, analysts face the challenge of quantifying indirect effects of the incentive payments, where indirect effects are defined as the changes in farming practices made elsewhere in response to the conservation incentive (Smith and Weinberg, 2004). For example, incentives that discourage some farmers from sequential years of corn production can increase the price of corn and increase corn production (and subsequent environmental impacts) elsewhere.

[16] Suitable data for analyses of hunter demands require population surveys with questions that relate to hunting opportunities and details on activities (see box "Estimating the Value of One More Duck").

[17] Reactive nitrogen is a general term used to represent the set of nitrogen compounds that support plant growth. Other chemicals include phosphorous, herbicides, and insecticides. Due to a paucity of data on the prevalence and impacts of these chemicals, these issues are left for future research.

[18] Other wetlands are known to remove nitrogen from surface waters, but the complexities of very similar wetlands' hydrology make it difficult to predict effects of similar wetlands. For example, in an analysis of depressional wetlands near the Chesapeake Bay, Denver et al. (2014) found that the depth of the aquifer below the wetland is the largest determinant of nitrogen removal. Information on aquifer depth is not readily available, making it difficult to predict the nitrogen removal rate of a restored wetland in some areas.

[19] The NatureServe data used in our analysis include all species that have a NatureServe Global Conservation Status Rank of G1or G2. A rank of G1 indicates that a species is critically imperiled across its entire range. A rank of G2 indicates the species is imperiled over its range. Rankings are based on various biological and external factors that may affect the long-term viability of a species (Regan et al., 2004). For details on species classifications, see appendix A, page 46 in www.natureserve.org/sites/default/files/ publications/files/natureserveconservationstatusfactors_apr12_1.pdf

[20] ERS analysis of NatureServe 2014 data (NatureServe, 2014).

[21] Counties where imperiled wetland-associated species are prevalent may not be important to agricultural policy if there are few opportunities to restore wetlands on agricultural land.

[22] Wetlands can reduce flood levels from short, heavy downpours when wetland water levels are relatively low, and they can capture a portion of the surge in water flows.

[23] Executive Order 12866 required agencies "to assess both the costs and the benefits of the intended regulation and, recognizing that some costs and benefits are difficult to quantify, propose or adopt a regulation only upon a reasoned determination that the benefits of the intended regulation justify its costs" (U.S. Government, 2010).

[24] These social cost of carbon (SCC) estimates are somewhat higher but not out of line with those reported elsewhere. For example, Nordhaus (2014) estimated the SCC to be $18.60 per ton in 2015 while the International Panel on Climate Change (IPCC) reported an SCC of $23.80. But Nordhaus (2014) found the SCC increase to be faster (3 percent annually), reaching $52.30 in 2050, compared with the IPCC estimate of $44.90.

[25] The market discount rate is appropriate because we are evaluating investments in wetland resources, not the social welfare effects of losses in lives and human welfare. We apply a 4-percent market rate of discount in this analysis, the longrun return on AAA bonds (U.S. Federal Reserve, 2012).

[26] Given that rural areas, almost by definition, have low population densities, open space is abundant.

[27] We were unable to include all costs, or all benefits, in our analysis so the estimated benefit-cost ratios may be too high or too low. However, for those benefits that were estimated, we expect that the spatial variations in benefit-cost ratios are realistic and informative for allocation decisions.

[28] Some of these counties may have no wetlands. Some of these counties may have few, if any, wetlands affected by agriculture. There is also the possibility that a wetland can have an indirect effect on a species that is not readily apparent. For example, a wetland's effect on nitrogen loadings could affect the survival rate of a species that is not wetland- associated.

[29] The soil-productivity benefit estimates reported in Hansen and Ribaudo (2008) should not be applied when land is not to be moved back into agricultural production. In cases where the land is idled temporarily, as with the CRP, one must account for the delay in the yield gains resulting from the increase in productivity.

[30] When market failures exist (such as when farmers do not pay for damages or are not compensated for benefits their practices generate), market equilibrium is not likely to be socially optimal.

[31] Because the quality of recreational fishing is likely to depend on both catch rates and the size of fish caught, developing a measure of such an amenity, $g_{FISHINGQUALITY}(f_{FISH}(x))$, is challenging.

[32] Values of *easting and northing* are expressed in meters and are based on the Universal Transverse Mercator, NAD 83, zone 14. In our analysis, easting and northing were both divided by 100,000.

[33] Strategic locations are locations where wetlands will capture surface and subsurface water flows from fields, as opposed to placing wetlands indiscriminately. When it comes to wetland location, the N removal estimates reported here should be viewed as upper-bound estimates.

INDEX

C

D

E

F

G

H

I

S

T

U

V

W

Y

Z